Advances in Ceramic Matrix Composites X

Technical Resources

Journal of the American Ceramic Society

www.ceramicjournal.org

With the highest impact factor of any ceramics-specific journal, the *Journal of the American Ceramic Society* is the world's leading source of published research in ceramics and related materials sciences.

Contents include ceramic processing science; electric and dielectic properties; mechanical, thermal and chemical properties; microstructure and phase equilibria; and much more.

Journal of the American Ceramic Society is abstracted/indexed in Chemical Abstracts, Ceramic Abstracts, Cambridge Scientific, ISI's Web of Science, Science Citation Index, Chemistry Citation Index, Materials Science Citation Index, Reaction Citation Index, Current Contents/ Physical, Chemical and Earth Sciences, Current Contents/Engineering, Computing and Technology, plus more.

View abstracts of all content from 1997 through the current issue at no charge at www.ceramicjournal.org. Subscribers receive full-text access to online content.

Published monthly in print and online. Annual subscription runs from January through December. ISSN 0002-7820

International Journal of Applied Ceramic Technology

www.ceramics.org/act

Launched in January 2004, *International Journal of Applied Ceramic Technology* is a must read for engineers, scientists,and companies using or exploring the use of engineered ceramics in product and commercial applications.

Led by an editorial board of experts from industry, government and universities, *International Journal of Applied Ceramic Technology* is a peer-reviewed publication that provides the latest information on fuel cells, nanotechnology, ceramic armor, thermal and environmental barrier coatings, functional materials, ceramic matrix composites, biomaterials, and other cutting-edge topics.

Go to www.ceramics.org/act to see the current issue's table of contents listing state-of-the-art coverage of important topics by internationally recognized leaders.

Published quarterly. Annual subscription runs from January through December.
ISSN 1546-542X

American Ceramic Society Bulletin

www.ceramicbulletin.org

The *American Ceramic Society Bulletin*, is a must-read publication devoted to current and emerging developments in materials, manufacturing processes, instrumentation, equipment, and systems impacting the global ceramics and glass industries.

The *Bulletin* is written primarily for key specifiers of products and services: researchers, engineers, other technical personnel and corporate managers involved in the research, development and manufacture of ceramic and glass products. Membership in The American Ceramic Society includes a subscription to the *Bulletin*, including online access.

Published monthly in print and online, the December issue includes the annual *ceramicSOURCE* company directory and buyer's guide. ISSN 0002-7812

Ceramic Engineering and Science Proceedings (CESP)

www.ceramics.org/cesp

Practical and effective solutions for manufacturing and processing issues are offered by industry experts. CESP includes five issues per year: Glass Problems, Whitewares & Materials, Advanced Ceramics and Composites, Porcelain Enamel. Annual subscription runs from January to December. ISSN 0196-6219

ACerS-NIST Phase Equilibria Diagrams CD-ROM Database Version 3.0

www.ceramics.org/phasecd

The ACerS-NIST Phase Equilibria Diagrams CD-ROM Database Version 3.0 contains more than 19,000 diagrams previously published in 20 phase volumes produced as part of the ACerS-NIST Phase Equilibria Diagrams Program: Volumes I through XIII; Annuals 91, 92 and 93; High Tc Superconductors I & II; Zirconium & Zirconia Systems; and Electronic Ceramics I. The CD-ROM includes full commentaries and interactive capabilities.

Advances in Ceramic Matrix Composites X

Ceramic Transactions Volume 165

*Proceedings of the 106th Annual Meeting
of The American Ceramic Society,
Indianapolis, Indiana, USA (2004)*

Editors

J.P. Singh
Narottam P. Bansal
Waltraud M. Kriven

Published by
The American Ceramic Society
PO Box 6136
Westerville, Ohio 43086-6136
www.ceramics.org

Please direct republication or special copying permission requests to the Staff Director, Technical Publications, The American Ceramic Society, PO Box 6136, Westerville, Ohio 43086-6136, USA.

For information on ordering titles published by The American Ceramic Society, or to request a publications catalog, please call 614-794-5890, or visit our website at www.ceramics.org

ISBN 1-57498-186-2

Contents

Mechanical Properties and Micromechanical Modeling

Preface

Ceramic composites are leading candidate materials for high-temperature structural applications. This proceedings volume contains papers given at a symposium on ceramic-matrix composites held during the 106th Annual Meeting and Exposition of the American Ceramic Society in Indianapolis, April 18-21, 2004. This symposium provided an international forum for scientists and engineers to discuss all aspects of ceramic composites. A total of 45 papers, including invited talks, oral presentations, and posters, were presented from 13 countries (the United States, Australia, Brazil, Cuba, France, India, Iran, Japan, Korea, New Zealand, Sweden, Taiwan, and Turkey). The speakers represented universities, industry, and government research laboratories.

This volume contains 15 invited and contributed peer-reviewed papers on various aspects of ceramic-matrix composites. The latest developments are covered in the areas of composite processing and characterization. The result is that all of the most important aspects necessary for understanding and further development of ceramic composites are discussed.

The organizers are grateful to all participants and session chairs for their time and effort, to authors for their timely submissions and revisions of the manuscripts, and to reviewers for their valuable comments and suggestions; without the contributions of all involved, this volume would not have been possible. Financial support from the American Ceramic Society is gratefully acknowledged. Thanks are due to the staff of the Meetings and Publications Department of the American Ceramic Society for their tireless efforts. Especially, we greatly appreciate the helpful assistance and cooperation of Greg Geiger throughout the production process of this volume.

Jitendra P. Singh
Narottam P. Bansal
Waltraud M. Kriven

Ceramic Fibers

SOL-GEL PROCESSING OF ALUMINA-ZIRCONIA FIBERS

J.Chandradass and M.Balasubramanian
Composites Technology Center & Dept. of Metallurgical & Materials Engineering
Indian Institute of Technology - Madras
Chennai - 600036
India.

ABSTRACT

Alumina-zirconia fibers were prepared by sol-gel method. The starting material used for the preparation of alumina sol was aluminium-tri-isopropoxide. Zirconia sol was prepared from zirconium oxychloride. Required quantity of zirconia sol was added to alumina sol, so that the final composite contains 5, 10, 15, and 20 wt.% zirconia. Suitable binder was added to the mixed sol and aged at room temperature for some gelation to occur. At the appropriate viscosity, fibers were formed in ammonia solution. The fibers were dried at room temperature and then sintered at 1600°C. X-ray diffraction analysis showed the presence of α-Al$_2$O$_3$ and t-ZrO$_2$ for fibers containing 5 wt.% zirconia and m-ZrO$_2$ was also present, when the zirconia content is 10 wt.% or more. Thermogravimetric analysis indicated the removal of most of the volatile up to 500°C. Differential thermal analysis and Fourier Transform infrared spectroscopy indicated the phase transition to α-Al$_2$O$_3$ in corroboration with X-ray studies. Tensile strength of the sintered samples was found to be highest for fibers containing 10 wt.% zirconia.

INTRODUCTION

The alumina-zirconia fiber is a potential candidate for the reinforcement of ceramic and metal matrix composites because of its excellent mechanical and physical properties such as strength and modulus and retention of these properties at high temperature [1, 2]. The development of alumina-zirconia fiber was based on the desire to improve upon the mechanical properties >99% α-alumina fiber, since it is often not useful for use in high temperature composites because of the deterioration in mechanical properties which result from grain growth and creep at high temperature. Alumina-zirconia fiber has been reported to improve high-temperature properties compared to alumina fiber [3]. The tensile strength of the fiber has been reported to be 2100 MPa. This value is about 50% higher than reported for Dupont's fiber, α-alumina [4]. The fine dispersion of tetragonal zirconia has been reported to contribute to the higher strength of alumina-zirconia fiber [5, 6]. Various methods for preparing ceramic fibers are melt spinning [7-10], melt extraction [11], pyrolysis of cured polymer fiber [12, 3], unidirectional freezing of gel [13, 14], and sol-gel processing [15-17]. The primary advantages of sol-gel processing over conventional melt routes are low temperature processing, homogeneity of product, uniform diameter of fiber, fine grain size and good control over final properties of the fiber [18-20]. Essentially sol-gel processing of ceramic fiber involve the following steps [21]

Preparation of the sol with suitable additive and achieving the right rheology for spinning.

Spinning of the solution to obtain gel fibers.

Calcinations of the gel fibers to obtain final oxide fibers.

It is important to control all of the above three stages in order to obtain high quality fiber with desired final properties. The present work aims at the preparation of alumina-zirconia fibers by sol-gel technique and the characterization of fibers.

EXPERIMENTAL PROCEDURE

Alumina sol was prepared according to the procedure described by Yoldas [22]. Aluminium-tri-isopropoxide (CDH, New Delhi) was dissolved in distilled water in a molar concentration of 1M refluxed at 80°C for 3 hours in the presence of acid catalyst (0.07 mole nitric acid) and then cooled.

Zirconium oxychloride (Otto Kemi, Mumbai) was dissolved in distilled water taken in a beaker to a molar concentration of 1M. Oxalic acid was taken in another beaker and dissolved in distilled water to a molar concentration of 1M. Both the solutions were mixed and stirred continuously until the solution becomes transparent [23].

Required amount of zirconia sol (5, 10, 15 or 20 wt. % Zirconia in the final composition) was mixed with alumina sol. Hydroxyl ethylcellulose (HEC) was added as binder to the sol to impart green strength and plasticity. Then the mixture was kept at room temperature for some gelation to occur. At the appropriate viscosity, sol was taken in a syringe and injected continuously to a gelation container (250ml beaker filled with 150ml of ammonia solution). The fiber was immediately transferred into drying pan and then dried at room temperature for 24 hours. After drying, the fibers have sufficient strength. Fibers were then sintered at 1600°C for 2 hours with a heating rate of 5°C/min.

The fibers were characterized using X-ray powder diffractometer with Cu-K$_\alpha$ radiation (SHIMADZU XD-DI), thermal analyzer (NETZSH STA409PC) and Fourier Transform Infrared spectrometer by the KBr method (PERKIN ELMER RXI). The tensile strength was determined by Universal tensile testing machine (Instron 4301). Fiber was mounted with adhesive on chart paper tabs for aligning and gripping. A 5mm gauge length and crosshead speed of 0.5mm/min was used in all these tests. The fracture load was converted to tensile strength by measuring the cross-sectional area of the fiber with an optical microscope. Ten samples were tested for each set and the average values are reported.

RESULTS AND DISCUSSIONS

The photographs of dried and sintered fibers of Al_2O_3-10 wt. % ZrO_2 are shown in Figs.1 & 2. The length and diameter of the fibers are varying from 1-4 cm and 100-200 μm respectively.

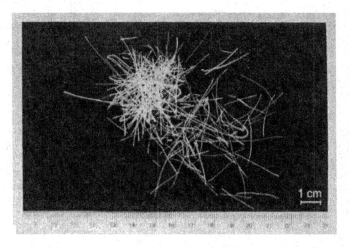

Fig.1 Photograph of as dried Al_2O_3-10 wt.% ZrO_2 fibers

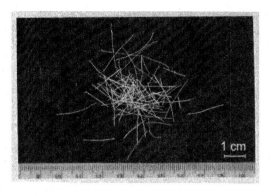

Fig.2 Photograph of sintered Al_2O_3-10 wt.% ZrO_2 fibers

Differential thermal analysis of alumina-zirconia fibers containing 5, 10, 15 and 20 wt.% zirconia was performed at 10°C/min and is shown in Fig.3. The curve has an endothermic peak around 105°C and an exothermic peak at 220°C indicate evolution of water and the decomposition of organic binder. The exothermic peak for the crystallization of α-alumina was not apparent, so a first order derivative was taken for 5, 10, 15 and 20 wt. % zirconia curves. It was found that the α-Al_2O_3 phase crystallization temperature was about 1290°C, which was higher, compared to the crystallization temperature reported by Yoldas [22]. The reason for this is that, during the transformation to α-alumina, the zirconium atoms come out of the cubic lattice sites and enters the interstitial and vacant sites causing expansion of the lattice along the a-axis. The zirconium atoms probably exert a dragging force on the diffusion of aluminium ions [23]. Because of these reasons, the transformation to α-Al_2O_3 occur at higher temperature in the presence of zirconia.

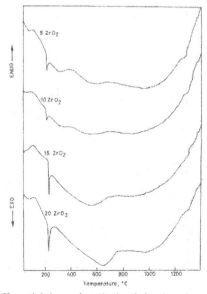

Fig.3 Differential thermal analysis of alumina-zirconia fiber

The TGA curves of alumina-zirconia fibers taken at a heating rate of 10°C/min are shown in Fig.4. A total weight loss of 61, 54, 52, 49% is observed for the fibers containing 5, 10, 15, and 20 wt. % zirconia respectively. The curve shows three stages of weight loss at different temperature, one up to 180°C and other between 180 and 250°C. The first two stages of weight loss are attributable to loss of adsorbed water and decomposition of organic binder as is evident from DTA curve. The weight loss between 250 and 400°C is due to loss of structural water (decomposition of structural –OH groups). Thus the TGA results show that the magnitude of weight loss is significantly different for samples containing different amount of zirconia.

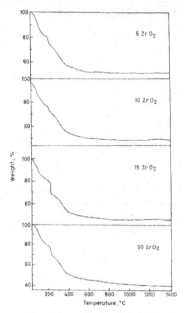

Fig.4 Thermogravimetric analysis of alumina-zirconia fibers

FTIR spectroscopic analysis for the dried gel and oxide fibers is shown in Figs.5 and 6. The FTIR results can be divided into four regions: (i) in the region between 3000 and 4000 cm^{-1} the broad band recorded at about 3480 cm^{-1} [24] in the gel fibers is due to the presence of molecular water. This peak is also present in oxide fibers with reduced intensity. This may be due to the absorption of moisture during testing; (ii) in the second region between 2000 and 3000 cm^{-1}, small peaks are noted at 2920 and 2850 cm^{-1} [25] that are due to the presence of Al-OH stretching vibration mode. These peaks are also present in oxide with reduced intensity; (iii) in the third region between 1000 and 2000 cm^{-1} broad peaks noted at 1650 cm^{-1} [25-27] is due to free as well as adsorbed water and the peaks observed at 1384 [25] and 1070 cm^{-1} [28] correspond to Al-OH bending mode. These peaks are also observed in oxide fibers with reduced intensity; (iv) in the fourth region between 1000 and 400 cm^{-1} peaks recorded are due to bonds involving Al and Zr. The peak at 480 cm^{-1} [29] corresponds to stretching vibration of Zr-O bonds. Peak at 610 cm^{-1} [23] observed in oxide fibers corresponds to AlO$_6$. Peak noticed at 540 cm^{-1} [29] is due to Zr-O stretching vibration. A peak observed around 450 cm^{-1} [30] indicate the absorption band of α-alumina.

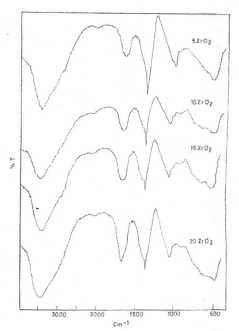

Fig.5 FTIR curves of as-dried alumina-zirconia fibers

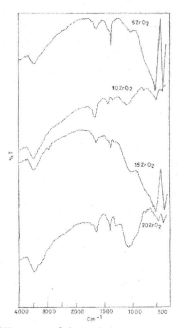

Fig.6 FTIR curves of sintered alumina-zirconia fibers

X-ray diffraction pattern of the dried and sintered alumina-zirconia fibers is shown in Figs.7 and 8. Very broad peaks of the as-dried alumina-zirconia fibers correspond to γ-AlOOH (boehmite) phase. The boehmite phase changes to α-Al$_2$O$_3$ and t-ZrO$_2$ alumina-zirconia fiber containing 5 wt.% zirconia after sintering at 1600°C for 2 hours whereas α-Al$_2$O$_3$, t-ZrO$_2$ and m-ZrO$_2$ is present for sample containing 10, 15 and 20 wt.% zirconia. Hence 100% t-ZrO$_2$ is retained in the 5 wt.% ZrO$_2$.

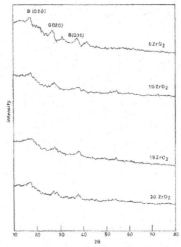

Fig.7 X-ray diffraction patterns of as-dried alumina-zirconia fibers

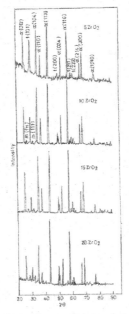

Fig.8 X-ray diffraction patterns of sintered alumina-zirconia fibers

It was reported that for lower ZrO_2 content, the zirconia grain size always lies below the critical diameter [31]. When the zirconia content is higher the grain size increases and part of the ZrO_2 is transformed into the monoclinic phase during cooling. Becher [32] considered that there was a critical volume fraction for a given particle size to cause transformation. It is based on the fact that an internal tensile stress is produced due to the thermal expansion mismatch of Al_2O_3 ($\alpha=8.1\times10^{-6}K^{-1}$) and ZrO_2 ($\alpha=10.5\times10^{-6}K^{-1}$). This internal tensile stress increases with increasing amount of ZrO_2. When zirconia content is above a critical level, the tensile stress is equal to the stress required for transformation. Thus zirconia grain growth and internal tensile stress are responsible for t→m transformation in samples containing 10 wt.% or higher amount of zirconia.

Fig.9 shows the surface morphology of alumina-zirconia fibers containing 5, 10 and 20 wt.% zirconia. The surface morphology of alumina-10 wt.% zirconia composite fibers consisted of alumina grains with ZrO_2 grains intertwined whereas alumina-zirconia fibers containing 5 and 20 wt.% zirconia shows interconnected porosity, open pores and closed pores with zirconia particle within the grains and along the grain boundaries.

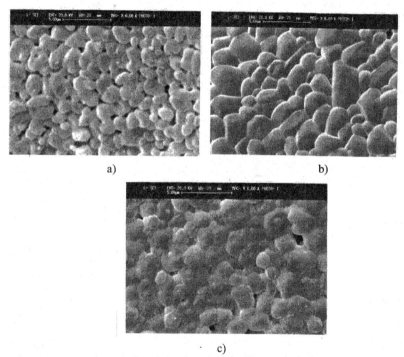

a)   b)

· c)
Fig.9 SEM micrographs of alumina-zirconia fiber surface a) Al_2O_3-5wt.% ZrO_2 ;
b) Al_2O_3-10wt.% ZrO_2 ; c) Al_2O_3-20wt.% ZrO_2

Fig.10 shows the fracture morphology of alumina-zirconia fiber containing 5, 10 and 20 wt.% zirconia. The fracture surface of all the samples shows intergranular type fracture.

The tensile strength measurement results are given in Table I. The tensile strength was found to be higher for Al_2O_3-10 wt.% ZrO_2. Al_2O_3-5 wt.% ZrO_2 contains only t-ZrO_2 phase, which are well below its critical size for transformation. So, this will not contribute for strength improvement. Al_2O_3-10 wt.% ZrO_2 contains t-ZrO_2 with small amount of m-ZrO_2.

The stress induced transformation can take place in this sample and that will improve the strength. Alumina-zirconia fibers with 15 and 20 wt.% ZrO_2 contain higher amount of m-ZrO_2. During load application, the contribution from stress induced transformation is less when there is less amount of t-ZrO_2. This is responsible for the lower strength values for 15 and 20 wt.% ZrO_2 samples.

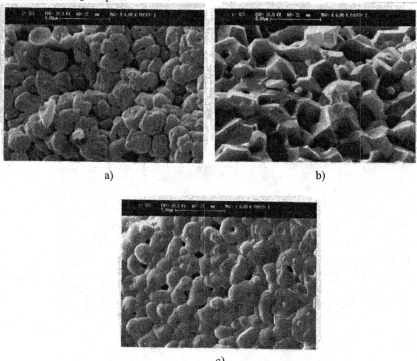

c)

Fig.10 SEM micrographs of alumina-zirconia fiber fracture surface a) Al_2O_3-5wt.% ZrO_2; b) Al_2O_3-10wt.% ZrO_2 ; c) Al_2O_3-20wt.% ZrO_2

Table I Tensile strength of alumina-zirconia fibers

S.No	Sample Name	Tensile Strength (MPa)
1.	Al_2O_3-5 wt.% ZrO_2	479 ± 40
2.	Al_2O_3-10 wt.% ZrO_2	846 ± 100
3.	Al_2O_3-15 wt.% ZrO_2	558 ± 40
4.	Al_2O_3-20 wt.% ZrO_2	522 ± 40

CONCLUSION

Alumina-zirconia fibers containing 5, 10, 15 and 20 wt.% zirconia was successfully fabricated by sol-gel processing. The phases present in sintered fibers are α-Al_2O_3 and t-ZrO_2 in alumina-zirconia fiber containing 5 wt.% ZrO_2 and α-Al_2O_3, t-ZrO_2 and m-ZrO_2 in alumina-zirconia fibers containing more than 5 wt.% zirconia. The tensile strength was found to be higher for Al₂O₃-10 wt.% ZrO₂ fibers.

ACKNOWLEDGEMENT
The authors are thankful to the Ministry of Human Resources and Development, Government of India for financial support to this work

REFERENCES
[1]X.Yang and R.J.Young, Determination of residual strains in Ceramic fiber reinforced composites using fluorescence spectroscopy, *Acta Mettallurgica et Materialia,* **43** [6] 2407-2416 (1995).

[2]X.Yang, X.Hu, R.J.Day, R.J.Young, Structure and deformation of high modulus alumina-zirconia fibers, *Journal of Materials Science,* **27** 1409-1416 (1992).

[3]S. Nourbakshi, F.L.Liang, H.Margolin, Characterization of Zirconia toughed alumina fiber, PRD-166, *Journal of Materials Science,* **8** 1252-54 (1989).

[4]T.Yoko, S.Kodama and H.Iwahara, Synthesis of polycrystalline alumina-zirconia using aluminum-zirconium precursor, *Journal of Materials Science,* **28** 105-110 (1993).

[5]D.J.Pysher, K.C.Goretta, R.S. Hodder Jr and R.E.Tressler, Strength of ceramics at elevated temperature, *Journal of the American Ceramic Society,* **72** [2] 284-288 (1989).

[6]V.Lavaste, M.H.Berger, A.R. Bunsell, J.Besson, Microstructure and Mechanical Characteristics of alpha-alumina-based fibers, *Journal of Materials Science,* **30** 4215-4225 (1995).

[7]F.T. Wallenberger, N.E.Weston, S.A.Dunn, Melt spun calcium aluminate fiber; Infrared transmission, *Journal of Non-crystalline Solids,* **124** 116-119 (1990).

[8]Yun-Mo-Sung, Stantley A.Dunn, Inviscid Melt-spun-high temperature alumina-magnesia fibers, *Journal of Materials Science,* **31** 3657-3660 (1996).

[9]Fernanado Fondeur, Brian S.Mitchel, Infra red studies of calcia-alumina fibers, *Journal of the American Ceramic Society,* **79** [9] 2469-73 (1996).

[10]Brian S.Mitchell, Infra red studies of calcia-alumina fibers formed via invisid melt Spinning, *Ceramic International,* **24** 67-71 (1998).

[11]M.Allahverdi, R.A.L.Drew, J.O.Strom-olsen, Melt extracted oxide ceramic fiber- the fundamentals, *Journal of Materials Science,* **31** 1035-1042 (1996).

[12]T.Yogo, H.Iwahara, Synthesis of α-alumina fiber from chelated aluminium alkoxide precursor, *Journal of Materials Science,* **26** 5292-5296 (1991).

[13]T.Maki, S.Sakka, Preparation of porous alumina fiber by unidirectional freezing of gel, *Journal of Materials Science Letters,* **5** 28-30 (1986).

[14]T.Maki, S.Sakka, Formation of alumina fibers by unidirectional freezing of gel *J.Non-cryst.Solids,* **82** 239-245 (1986).

[15]T.Makki, S.Sakka, Preparation of alumina fiber by sol-gel method, *Journal of Non-crystalline Solids,* **100** 303-308 (1988).

[16]A.Towata, H.J.Hwang, M.Yaswoka, M.sando, K.Niihara, Preparation of polycrystalline YAG/alumina composites fibers and YAG fibers by sol-gel method, *Composites: Part A* **32,**1127-1131 (2001).

[17]Y.H.Chiou, M.T.Tsai, H.C.Shih, The preparation of alumina fiber by sol-gel processing, *Journal of Materials Science,* **29** 2378-2388 (1994).

[18]H.Schmidt, Chemistry of material preparation by the sol-gel process, *Journal of Non-crystalline Solids,* **100** 51-64 (1988).

[19]J.D.Mackenzie, Application of the sol-gel process, *Journal of Non-crystalline Solids,* **48** 162-168 (1988)

[20]S.Sakka, Sol-gel glasses and their future applications, *Transactions of the Indian ceramic society*, **46** [1] 1-11 (1987).

[21]R.Venkatesh, S.R. Ramanan, Influence of processing variables on the microstructure of sol-gel spun alumina fiber, *Materials Letters*, **55** 189-195 (2002).

[22]B.E.Yoldas, Alumina sol preparation from alkoxides, *American Ceramic Society Bulletin*, **54** 289-290 (1975).

[23]M.Balasubramanian, Ph.D. Thesis, Processing of alumina-zirconia powders and Composites, IIT Madras, May 1996,.

[24]R.Venkatesh, S.R.Ramanan, Effect of organic additives on the properties of sol-gel spun alumina fiber, *Journal of the European Ceramic Society*, **20** 2543-49 (2000).

[25]N.S.Lakshmi, Ph.D. Thesis, Studies on process development and property evaluation alumina minispheres and fibers prepared from boehmite sol, Anna University, Chennai, December 2001.

[26]Tao He, Xiuling Jiao, Dairong Chen, Mengkai Lu, Duorang Yuan, Dong Xu, Synthesis of Zirconia sols and fibers by electrolysis of zirconium oxychloride, *Journal of Non-crystalline Solids*, **283** 56-62 (2001).

[27]Ph.Colomban, Structure of oxide gels and glasses by Infrared and Raman scattering, *Journal of Materials Science*, **24** 3002-3010 (1989).

[28]M.Low, R.McPherson, Crystallization of gel derived alumina and alumina-zirconia ceramics, *Journal of Materials Science*, **24** 892-898 (1989).

[29]T.Y.Tsing, C.C.Lin, J.T.Liaw, Phase transformation of gel derived magnesia partially stabilized zirconia, *Journal of Materials Science*, **22** 965-972 (1987).

[30]T.Assih, A.Ayral., M.Abenoza, J.Phalippou, Raman study of alumina gels, *Journal of Materials Science*, **23** 3326-3331 (1988).

[31]J.P.Bach, F.Thevenot, Fabrication and Characterization of zirconia toughened Alumina obtained by inorganic and organic precursors, *Journal of Materials Science*, **24** 2711-27 (1989).

[32]P.F.Becher, Toughening behaviour in ceramics associated with the transformation of tetragonal ZrO_2, *Acta Mettallurgica et Materialia,* **34** 1885-1891 (1986).

Oxide and Non-Oxide
Ceramics and Composites

FABRICATION OF TRANSPARENT POLYCRYSTALLINE SILICON NITRIDE CERAMIC

R.-J. Sung, T. Kusunose, T. Nakayama, T. Sekino
The Institute of Scientific and Industrial Research
Osaka University, Ibaraki
Osaka, 562-0047 Japan

S.-W. Lee
Department of Materials Engineering
Sunmoon University, Asan
Chungnam, 336-708 Korea

K. Niihara
The Institute of Scientific and Industrial Research
Osaka University, Ibaraki
Osaka, 562-0047 Japan

ABSTRACT

As hot pressing silicon nitride, sintering additives such as MgO, Al_2O_3 and Y_2O_3 are generally used. The use of sintering additives changes the properties of hot pressed silicon nitride such as mechanical, thermal and optical. In the present paper, silicon nitride was hot pressed with various amount of AlN as a sintering additive. Mechanical properties such as density, hardness, Young's modulus, fracture toughness and flexural strength of hot pressed silicon nitride were measured with regards to the content of AlN. Also microstructure was observed using a SEM. The effect of α/β phase fraction on the mechanical and optical properties of silicon nitride was investigated as a function of the amount of AlN additive. Transparent polycrystalline silicon nitride consists of 75 vol.% α-phase Si_3N_4 and 25 vol.% β-phase Si_3N_4. The elongation of β-Si_3N_4 was suppressed. The resulting polycrystalline silicon nitride showed superior properties such as about 64% transmittance at wavelength of 2.5 μm with 300 μm thickness.

INTRODUCTION

Silicon nitride is one of the most attractive engineering ceramic materials for high-temperature applications because of its superior oxidation, thermal shock resistance and mechanical strength. Since silicon nitride is difficult to be fully densified due to their strong covalent bonding, it is commonly densified through liquid phase sintering by adding sintering additives such as MgO, Al_2O_3, Y_2O_3 and Al_2O_3-Y_2O_3. A kind and amount of composition of the additives are not only of decisive influence on the sintering parameters (temperature, pressure, time, atmosphere), but also on the resulting phase relations and microstructures, which emphatically determine many properties of Si_3N_4 ceramics.

The additives form a liquid phase by the reaction with Si_3N_4 and surface SiO_2 on Si_3N_4 particles to promote densification.[1] However, the type of additives results in different conductivities, because sintering additives influenced significantly the phase transformation from α- to β-Si$_3$N$_4$ and the formation of crystal defects. In this study, MgO and AlN were added as sintering additives into silicon nitride. The mechanical and optical properties of silicon nitride ceramics were closely related to the amount of α-phase of hot pressed silicon nitride.

Polycrystalline ceramics became available for optical applications in the early 1960s when Coble first made translucent Al$_2$O$_3$.[2] Since then, a number of oxide and nitride ceramics have been developed for optical and other applications.[3-9] Transparency is one of optical properties of materials. A glass is transparent because it has short-term ordered structure only, and also it is optically isotropic. There is no grain boundary in a glass, and hence very little scattering or absorption of light. Ceramics are generally polycrystalline. The grain boundaries in ceramics strongly scatter light. However, if the grain size is smaller than the wavelength of the visible light (0.4-0.7 μm), light can transmit through the ceramic just like it travels through a grating.

From viewpoint of optical properties, oxide ceramic is commonly used because they have large optical bandgap and hence transparent polycrystalline ceramic is fairly and easily obtained. In the case of non-oxide ceramic such as silicon nitride, it is usually opaque. Owing to the difference in light absorption, the impurity phases in a ceramic will certainly affect transparency, usually decreasing transparency by scattering or absorption. Porosity also influences transparency in the same manner. In short, density, purity, and grain size are the key factors that influence the transparency of a ceramic. To achieve transparency in a ceramic, efforts should be made to eliminate or minimize scattering or absorption of light.

EXPERIMENTAL PROCEDURES

High purity α-Si$_3$N$_4$ powder (>99.5% pure, >95% α-phase content, 0.2-μm average particle size, SN-E10, Ube Co., Japan) was used as a starting material. It was mixed with 3 wt.% MgO (>99.9% pure, <5ppm carbon, zinc and iron, Soekawa Chem., Japan) and various amount (0, 1, 3, 6, 9 wt.%) of AlN (>99.99% pure, <0.8% oxygen content, Grade F, Tokuyama Co., Japan). These powders were packed into the carbon mold and were sintered by hot-pressing in the nitrogen atmosphere under the applied pressure of 30MPa. Hot pressing temperature was 1900°C, and holding time was 1 hour. The density of hot pressed silicon nitride was measured by using Archimedes Method with toluene. Volume fractions of α- or β-phase of Si$_3$N$_4$ were determined by X-ray diffraction (XRD, RU-200B, RIGAKU Co., Ltd., Japan) with a CuKα radiation operated at 50kV and 150 mA. Young's modulus was measured by resonance method after carbon coating. The hardness was obtained by using Zwick 3212 hardness tester with Vickers indentor. Indentation was performed for 15 seconds at a load of 98N. Fracture toughness was determined by indentation fracture method at a load of 98N. Flexual strength was measured by three point bending test with a 25mm span under a cross- head speed of 0.5 mm/min at room temperature.

The polished surface was etched using a commercial plasma etching apparatus. The microstructures were examined with scanning electron microscope (SEM, S-500, Hitachi Co., Ltd., Japan). The transmittance was measure by UV spectra in the measuring range from 200nm to 2500nm.

RESULTS AND DISCUSSION

The relative density of hot pressed silicon nitride was over 99%, and it could be sintered with the high density. α-Si$_3$N$_4$ and β-Si$_3$N$_4$ were determined by X-ray diffraction analysis of the hot pressed silicon nitride as shown in Fig. 1. The volume fraction of α-phase Si$_3$N$_4$ was determined by the method of Gazzara and Messier[10] and shown in Table I. It is found that the volume fraction of α phase increased with increasing the amount of AlN. Also it was very difficult to search the peaks of MgO and AlN after hot pressing because they are dissolved in the grain and grain boundary. In the case of 0 wt.% content of AlN, β-phase Si$_3$N$_4$ only was found. It means that α-phase Si$_3$N$_4$ perfectly transformed into β-phase Si$_3$N$_4$.

Table I. Volume fraction of α-/β-phase of hot pressed silicon nitride with various AlN additives

(Vol.%)

	α-phase	β-phase
0 AlN	0	100
1 AlN	0	100
3 AlN	18	82
6 AlN	63	37
9 AlN	75	25

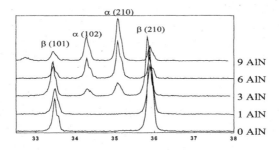

Fig. 1 X-ray diffraction patterns of hot pressed silicon nitride with various content of AlN. α-phase Si$_3$N$_4$ peaks appeared from 3 wt.% of AlN. The volume fraction of β-phase Si$_3$N$_4$ decreased with increasing the amount of AlN. At 9 wt.% AlN, the volume fraction of α-phase Si$_3$N$_4$ is much more than that of β-phase Si$_3$N$_4$.

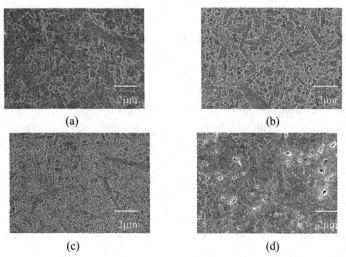

Fig. 2 SEM micrographs of the etched surface for hot pressed silicon nitride
with various content of AlN : (a) 0 wt.%, (b) 1 wt.%, (c) 3 wt.% and (d) 9 wt.%.

Fig. 2 shows SEM micrographs of the etched surface of silicon nitride depending on the amount of AlN. The large elongated grain of β-Si$_3$N$_4$ decreased with increasing the amount of AlN. In the case of 9 wt.% content of AlN, it has many pores comparing with other contents.

Fig. 3 SEM micrographs of fractured surfaces of silicon nitride:
(a) 0 wt.% AlN, and (b) 9 wt.% AlN.

Fig. 3 (a) shows the fractured surface of silicon nitride, which has heterogeneous microstructure including 100 vol.% β-Si$_3$N$_4$. Silicon nitride with 9wt.% of AlN has 25 vol.% β-Si$_3$N$_4$ but the shapes of grains were similar to α-Si$_3$N$_4$ as shown in fig. 3 (b). Similar to β-Si$_3$N$_4$ the α-Si$_3$N$_4$ is produced by liquid-phase sintering. This reaction starts at temperatures above 1450°C.[11] Amount of liquid available for densification is quickly reduced due to the formation of α-phase.

Hardness and Young's modulus were influenced depending on the volume fraction of α-phase Si₃N₄ and glassy phase at the grain boundary. The Young's modulus of silicon nitride increased with increasing the volume fraction of α-phase. Hardness increased with increasing the volume fraction of α-phase as shown in Fig. 4. As the reflection of the high hardness of α-phase Si₃N₄, Vickers hardness of hot pressed Si₃N₄ decreased with increasing the amount of AlN.

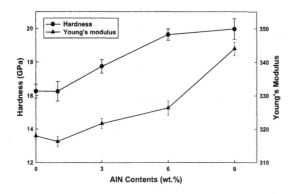

Fig. 4 Comparison of hardness and young's modulus of hot pressed silicon nitride with various content of AlN.

Flexual strength as well as fracture toughness decreased with decreasing the volume fraction o f β-phase Si₃N₄ as shown in Fig. 5. In this study, we tried to improve the thermal conductivity of silicon nitride as well as flexural strength by adding AlN into Si₃N₄, but flexural strength and fracture toughness decreased with increasing the amount of AlN.

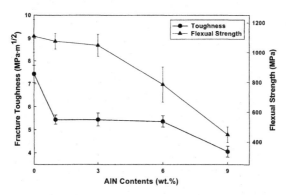

Fig. 5 Comparison of toughness and flexual strength of hot pressed silicon nitride with various content of AlN.

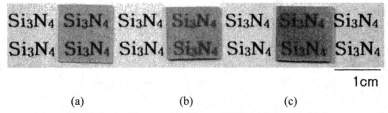

1cm

(a)   (b)   (c)

Fig. 6 Optical images of silicon nitride with 9wt.% AlN

depending on the thickness: (a)300μm, (b) 500 μm, and (c) 600 μm.

Fig. 6 shows optical images of silicon nitride with 9 wt.% AlN depending on the thickness, (a) 300μm, (b) 500μm, and (c) 600μm. We found that transparent polycrystalline silicon nitride can be sintered with 3 wt.% MgO and 9 wt.% AlN as shown in fig 6.

Microstructure of transparent polycrystalline silicon nitride depends strongly on the grain shape and size distribution. It has a highly dense structure, with small, uniform 75% α- and 25% β-phase grains and pores. Fig. 7 shows transmittance of transparent silicon nitride depending on thickness: (a) 300μm, (b) 600μm, (c) 1mm. The maximum transmittance, 64%, is observed at 2.5 μm in the infrared region. The transmittance decreased with increasing the thickness of the sample because of increasing of heterogeneousness of β-phase.

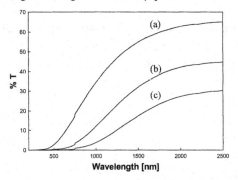

Fig 7. Transmittance of transparent silicon nitride depending on thickness:
(a) 300μm, (b) 600μm, and (c) 1mm.

CONCLUSIONS

Silicon nitride was hot pressed with 3 wt.% MgO and various amount of AlN. The relative density of hot Pressed Si_3N_4 was over 99%. Transparent polycrystalline silicon nitride was successfully fabricated by hot press sintering method at 1900°C with adding 3 wt.% MgO and 9 wt.% AlN as sintering additives. Regarding to the volume fraction of α phase in silicon nitride, mechanical and optical properties were influenced as followed:

(1) The hardness and Young's modulus increased with increasing the volume fraction of α-phase fraction as the reflection of higher hardness of α-phase Si_3N_4.

(2) The fracture toughness decreased with decreasing the amount of the elongated β-phase. And flexual strength decreased with increasing the volume fraction of α-phase Si_3N_4 as well as the amount of glass phase at the grain boundary.

(3) The maximum transmittance, 64%, is observed at 2.5 μm in the infrared region. Transparent polycrystalline silicon nitride consists of 75 vol.% α-phase Si_3N_4 and 25 vol.% β-phase Si_3N_4.

REFERENCES

[1]M. Mitomo, "Pressure sintering of Si_3N_4," *Journal of Materials Science*, **11** 1103-1107 (1976).

[2]R.L. Coble, "Transparent Alumina and Method of Preparation," U.S. Pat. No. 3 026 210, Mar. 20, 1962.

[3]R.J. Bratton, "Translucent Sintered $MgAl_2O_4$," *Journal of the American Ceramic Society*, **57**[7] 283-285 (1974).

[4]J.W. McCauley and N.D. Corbin, "Phase Relations and Reaction Sintering of Transparent Cubic Aluminum Oxynitride Spinel (ALON)," *Journal of the American Ceramic Society*, **62**[9-10] 476-479 (1979).

[5]N. Kuramoto, H. Taniguchi, I. Aso, "Development of Translucent Aluminum Nitride Ceramics", *American Ceramic Society Bulletin*, **68**[4] 883-887 (1989).

[6]M. Shimada, T. Endo, T. Saito, T. Sato, "Fabrication of Transparent Spinel Polycrystalline Materials," *Materials Letters*, **28** 413-415 (1996).

[7]A. Granon, P. Goeuriot, F. Thevenot, "Aluminum Magnesium Oxynitride: A New Transparent Spinel Ceramic," *Journal of the European Ceramic Society*, **15** 249-254 (1995).

[8]J.-G. Li, T. Ikegami, J.-H. Lee, T. Mori, "Low-Temperature Fabrication of Transparent Yttrium Aluminum Garnet (YAG) Ceramics without Additives," *Journal of the American Ceramic Society*, **83**[4] 961-963 (2000).

[9]R. Apetz and Michel P. B. van Bruggen, "Transparent Alumina: A Light-Scattering Model," *Journal of the American Ceramic Society*, **86**[3] 480-486 (2003).

[10]C. P. Gazzara and D. R. Messier, "Determination of Phase Content of Si_3N_4 by X-ray Diffraction Analysis," *American Ceramic Society Bulletin*, **56**[9] 777-780 (1977).

[11]M. Menon and I.W. Chen, "Reaction Densification of α'-SiAlON: I, Wetting Behavior and Acid-Base Reactions," *Journal of the American Ceramic Society*, **78**[3] 545-553 (1995).

A MODEL CERIUM OXIDE MATRIX COMPOSITE REINFORCED WITH A HOMOGENEOUS DISPERSION OF SILVER PARTICULATE - PREPARED USING THE GLYCINE-NITRATE PROCESS

K. Scott Weil and John S. Hardy
Pacific Northwest National Laboratory
Richland, WA 99352

ABSTRACT

Recently a new method of ceramic brazing has been developed. Based on a two-phase liquid composed of silver and copper oxide, brazing is conducted directly in air without the need of an inert cover gas or the use of surface reactive fluxes. Because the braze displays excellent wetting characteristics on a number ceramic surfaces, including alumina, various perovskites, zirconia, and ceria, we were interested in investigating whether a metal-reinforced ceramic matrix composite (CMC) could be developed with this material. In the present study, two sets of homogeneously mixed silver/copper oxide/ceria powders were synthesized using a combustion synthesis technique. The powders were compacted and heat treated in air above the liquidus temperature for the chosen Ag-CuO composition. Metallographic analysis indicates that the resulting composite microstructures are extremely uniform with respect to both the size of the metallic reinforcement as well as its spatial distribution within the ceramic matrix. The size, morphology, and spacing of the metal particulate in the densified composite appears to be dependent on the original size and the structure of the starting combustion synthesized powders.

INTRODUCTION

Improvements in lightweighting the materials used in freight truck and automotive engine and body components translate directly into increased fuel utilization and reduced emissions. One of the most promising families of materials in this regard are the discontinuously reinforced composites (DRC), including metal-matrix and ceramic-matrix types. Relative to the metal-matrix variety, ceramic materials offer both higher weight-specific and high temperature mechanical properties, as well as exhibit excellent wear and corrosion resistance; key considerations in designing gas and diesel engine components such as piston heads, cylinder liners, valve heads and stems, and direct injection ports. In addition, the use of particulate reinforcement within the ceramic can potentially mitigate one of the fundamental concerns with the corresponding monolithic material: low fracture toughness [1]. In this way, the inherent fracture toughness of the matrix is increased by one or more mechanisms including: bridging, deflection, and/or blunting of the fracture initiating crack front [2 - 4]. Specifically, metallic reinforcement materials offer a means of not only improving fracture toughness, but also keeping manufacturing costs low.

A primary issue in developing any discontinuous reinforced composite, whether it be metal- or ceramic-based, is the dispersivity of the reinforcement phase - i.e. is it uniform? Composites prepared by the traditional method of mechanical mixing are typically plagued by particulate clustering, which is due to the segregation of the dissimilar powders during mixing and compaction. To overcome this, we elected to form the composite directly in the powder synthesis step. We adopted a combustion-based approach, developed at PNNL [5] in the early 1990's, to prepare a multi-phase powder that when consolidated and appropriately heat treated would form the composite material of interest. Referred to as the glycine-nitrate combustion synthesis process, this technique consists of two basic steps: the formation of an aqueous metal nitrate-glycine solution and the subsequent heating of this solution to dryness and eventual autoignition, at which point a self-sustaining combustion reaction takes place that produces the final powder product. In this process, the glycine serves two purposes: (1) it prevents precipitation of the metal salts as the water is evaporated, thereby ensuring that the metal ions remain molecularly mixed in solution up to the point of combustion or chemical conversion, and (2) it acts as the fuel for the combustion reaction by undergoing oxidation with the nitrate ions during heating as shown in the example reaction [1] below:

$$12 \, Fe(NO_3)_3 \; + \; 8 \, H_2NCH_2CO_2H \; \Rightarrow \; 6 \, Fe_2O_3 \; + \; 20 \, H_2O \; + \; 16 \, CO_2 \; + \; 13 \, N_2 \qquad [1]$$

metal nitrate   amino acid, glycine   metal oxide
(oxidizer)   (fuel)   (product)

What is unique in our selection of materials is that we chose to study a new braze composition, Ag-CuO, as the metallic reinforcement phase. Recently developed as an alternative method of ceramic-to-metal brazing specifically for fabricating high temperature solid-state devices such as oxygen generators, Ag-CuO displays excellent wettability with a number ceramics, including alumina, various perovskites, zirconia, and ceria [6]. The braze is designed for both processing and application in oxidizing environments and does not requires the use of an inert gas cover or surface active fluxing agents during melting. Here we report on the synthesis of the powders and the initial results from powder compaction and densification.

EXPERIMENTAL
Materials and Processing
Cerium oxide and a braze composition of 4mol% CuO in Ag were chosen as the model system on which to conduct our investigation. Neither silver, copper metal, or CuO display any reactivity with CeO_2, making it an ideal inert ceramic matrix material in which to construct a series of model ductile metal-reinforced CMCs. Stock silver, copper, and cerium nitrate solutions were prepared in 1M concentration by dissolving the respective reagent-grade salts [$AgNO_3$, $Cu(NO_3)_2 \cdot 6H_2O$ and $Ce(NO_3)_3 \cdot 6H_2O$, Alfa Aesar] in de-ionized water. The cation concentration of each solution was adjusted to match the target within ±0.5%, as determined by EDTA titration. The nitrate solutions were then mixed by weight in the ratios necessary to meet the desired compositions of the final as-prepared composites. In each case, the ratio of silver nitrate to copper nitrate was kept constant at the value required to achieve the target composition of 4mol% CuO in Ag within the composite. A stoichiometric amount of glycine powder (99.5%, Alfa Aesar), as typified by Reaction [1] above, was then dissolved into each of the mixed aqueous solutions, after which a 40ml aliquot was drawn and heated in a 4L stainless steel beaker on a hot plate. The solution was allowed to boil to dryness, at which point it autoignited or spontaneously "burned". A fine stainless steel mesh was used to cover the beaker and retain the exothermically synthesized product within.

Once prepared, the powders were uniaxially pressed under 10ksi of pressure into 20mm diameter pellets that were further densified by cold isostatic pressing at a hydrostatic pressure of 20ksi. Prior to sintering, the average density of the pellets was estimated to be ~70% of theoretical, based on geometric measurements and a rule of mixtures calculation. The pellets were subsequently sintered in air under the following thermal cycle: heat at a rate of 10°C/min to 1000°C, isothermally hold at this temperature for 2hrs, then cool at 5°C/min to room temperature. Because the solidus temperature of Ag-4mol% CuO is 958°C, this composition was expected to form a liquid phase during sintering.

Characterization

The as-synthesized composite powders were characterized by particle size analysis (PSA), X-ray diffraction (XRD), scanning and transmission electron microscopy (SEM and TEM), and energy dispersive X-ray analysis (EDX). PSA was performed to determine powder agglomerate size using a Microtrac Model S3000 Particle Size Analyzer. X-ray diffraction (XRD) analysis was carried out with a Philips Wide-Range Vertical Goniometer and a Philips XRG3100 X-ray Generator over a scan range of 20–80° 2Θ with a 0.04° step size and 2s hold time. XRD pattern analysis was conducted using Jade 6+ (EasyQuant) software. All of the XRD specimens were prepared using the same specimen holder and preparation technique to ensure sample consistency. SEM analysis of the powders and cross-sectioned CMC compacts was conducted using a JEOL JSM-5900LV equipped with an Oxford Energy Dispersive X-ray Spectrometer (EDS) system that employs a windowless detector for quantitative detection of both light and heavy elements, while more detailed microstructural analysis was completed in a JEOL 2010F Field-emission gun Transmission Electron Microscope (FEG-TEM). On the TEM, an Oxford EDS was also used for compositional measurements. A 0.7 nm diameter probe allowed compositional information to be obtained on single particles within an agglomerate.

RESULTS AND DISCUSSION

Two target compositions were chosen for the initial set of experiments reported here: CeO_2 reinforced with either 5 or 10vol% of the Ag-4mol% CuO phase, respectively referred to as CeO_2/5 Ag-CuO and CeO_2/10 Ag-CuO. XRD analysis demonstrated that in both cases the only phases formed upon exothermic combustion were Ag, CuO, and CeO_2. The average agglomerate size for each set of powder was 239nm for CeO_2/5 Ag-CuO and 153nm for CeO_2/10 Ag-CuO. In a previous investigation on the combustion synthesis of Cu/CeO_2 precursor powders, we found that stoichiometric glycine-to-nitrate ratios gave us a similar trend: finer aggolmerate size with decreasing ceria content [7]. We speculated that this was due to a lower combustion temperature exhibited by precursor compositions of lower cerium nitrate content; i.e. the the $Cu(NO_3)_2 \Rightarrow$ CuO reaction reaction is significantly less exothermic than the $Ce(NO_3)_3 \Rightarrow CeO_2$. We presume a similar mechanism occurs here in the case of the silver-copper-cerium nitrate GNP precursors.

Shown in Figure 1 are TEM micrographs of the typical agglomerate structures observed in the two as-combusted powder products. In both cases, we observed that the crystallites of the three phases appeared to be uniformly dispersed within the agglomerates. In general the size of the crystallites was on the order of 5 - 30nm, with the ceria crystallites consistently observed to be slightly larger in size than either the silver or copper oxide. Obvious from the micrographs is an apparent trend in crystallite size with composition, i.e. a reduced crystallite size with decreasing ceria content. Again as found in the Cu/CeO_2 precursor study, crystallite size is also dependent on the composition of the nitrate solution, as well as on the glycine-to-nitrate ratio. Both variables can have a significant effect on the exothermic combustion temperature of the reaction, which can influence the crystallite size in the final powder [5].

Upon compaction and sintering in air at 1000°C for 2hrs, the densities of the two composites were found by the Archimedes method to be greater than 90% of theoretical density (TD). The density of the composite containing the higher reinforcement content was slightly higher than that of the second composite body, with 91% and 94% TD measured respectively for $CeO_2/5$ Ag-CuO and $CeO_2/10$ Ag-CuO. Back scattered electron images of the two densified composites in cross-section are shown in the micrographs of Figures 2 and 3. EDX analyses of both specimens confirm that the matrix (the gray regions in the micrographs) is essentially pure CeO_2, containing no copper oxide or silver. The CuO is always found in conjunction with the silver (appearing as the lighter phase in the micrographs). As was anticipated from the sub-theoretical sintered densities, a measurable amount of porosity can also be seen in each composite.

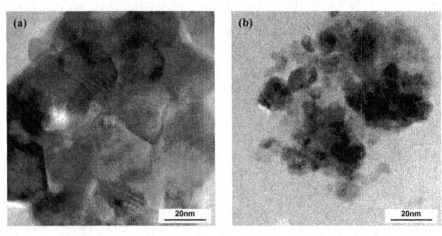

Figure 1   TEM micrographs of typical agglomerates in the two combustion synthesized composite powders: (a) $CeO_2/5$ Ag-CuO and (b) $CeO_2/10$ Ag-CuO.

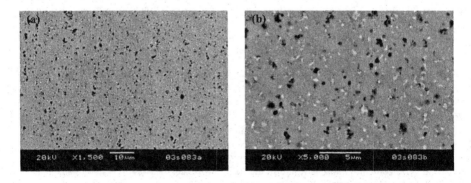

Figure 2   Cross-sectional SEM micrographs of a 5vol% Ag-4CuO in CeO_2 composite (specimen $CeO_2/5$ Ag-CuO): (a) 1500x and (b) 5000x magnification. The light phase is the Ag-4mol% CuO reinforcement phase, the black regions are porosity, and the gray regions are ceria.

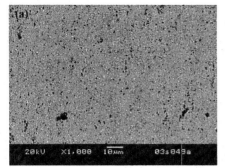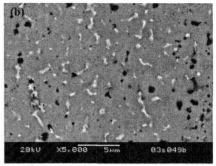

Figure 3   Cross-sectional SEM micrographs of a 10vol% Ag-4CuO in CeO₂ composite (specimen CeO₂/10 Ag-CuO): (a) 1500x and (b) 5000x magnification. The light phase is the Ag-4mol% CuO reinforcement phase, the black regions are porosity, and the gray regions are ceria.

It is interesting to note that to a large extent, the spatial homogeneity of phases observed in the powders is retained in the microstuctures of the densified composites, with an average interparticle spacing of 1.3μm, as estimated by the line-intercept method, in the CeO₂/5 Ag-CuO sample and 2.4μm in the CeO₂/10 Ag-CuO specimen. In both materials, the metallic phase assumes an exaggerated angular morphology. In the CeO₂/5 Ag-CuO specimen, the size ofthe silver particulate ranges from sub-micron size (on the order of 100nm) to as large as 1.5μm, with an average size of 0.6μm, whereas in the CeO₂/10 Ag-CuO specimen a somewhat larger reinforcement size is observed, on average ~ 1.3μm. The size of the porosity appears to display the opposite trend, with an average diameter of 0.9μm measured in the CeO₂/5 Ag-CuO sample and 0.7μm found in the CeO₂/10 Ag-CuO composite. Although it was not directly observed in these specimens, it has been previously reported that upon cooling below the solidus temperature for Ag-4mol% CuO the CuO precipitates out of liquid solution along the substrate/braze interface [6]. Thus, we expect that under high enough magnification, the majority of the CuO would be observed along the interface between the silver particles and the ceria matrix.

The morphologies of the porosity and metallic particles are consistent between the two samples and suggestive of the likely mechanisms for their origins. For example, based on the rounded shapes of the pores in both specimens, we suspect that they are intragranular in nature, having formed from the surface driven consolidation of smaller intercrystallite pores that originally arose early in the sintering process. The angular morphology of the silver particulate is indicative of an intergranular structure that formed as the original powder agglomerates began to neck. Given the low sintering temperature* and the relatively high final densities of the two specimens, it is likely that ceria sintering was enhanced by the liquid Ag-CuO phase. A proposed mechanism for the evolution and formation of the microstructures in these composites is shown schematically in Figure 4. The following steps are speculated to occur as the microstructure evolves from the compacted GNP powders to the final densified composite:

1. Step 1 (corresponding with ① in Figure 4) – the microstructure of the composite GNP powder after uniaxial and isostatic compaction. The crystallite structure of one

---

* Typical sintering conditions required to achieve 90+% density in CeO₂ are 1250°C in air for 4hrs [8].

agglomerate is shown, with the CeO_2 designated by the white speheres, the CuO by light gray, and the silver by dark gray. For the sake of clarity, the other three agglomerates are indicated by outline. Note the occurrence of porosity between the crystallites that make up each agglomerate. We believe these are responsible for the fine structure found in the final densified specimens.

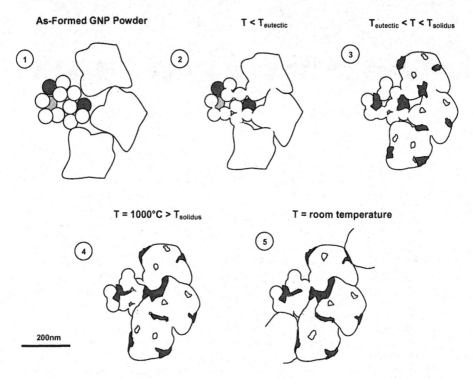

Figure 4   A schematic of the proposed mechanism of microstructural evolution in the Ag-CuO/CeO$_2$ composites fabricated in this study. Note that in illustrations 1 and 2 the microstructural details of the three "white" agglomerates are not shown for the sake of clarity.

2. Step 2 – in which the temperature is less than the eutectic temperature of Ag-4mol% CuO (932°C). Initial necking begins to take place primarily between the CeO$_2$ crystallites within each agglomerate, causing the agglomerates to shrink in size. Weak neck formation also likely takes place between crystallites in neighboring agglomerates.

3. Step 3 – in which the temperature is greater than the eutectic temperature, but less than the solidus temperature of Ag/4vol% CuO (958°C). For purposes of illustration, the structures of the Ag-CuO and pores have been included in the three

previously outlined agglomerates. At this temperature, a liquid forms where silver and CuO contact. Because of the wettability of this liquid on ceria (the contact angle is ~63° [9]), the surface tension of the molten Ag-CuO begins to pull the agglomerates closer together, enhancing the degree of contact and therefore sintering between the agglomerates. Voids between the original crystallites have become pores trapped within the former agglomerate structure.

4. Step 4 – in which the temperature is greater than the solidus temperature of Ag/4vol% CuO. The Ag-CuO is fully molten and begins to pool in crevices formed by the thickening necks between the former agglomerates, thus giving the metallic phases their characteristic morphology observed in Figures 2 and 3. Grain growth initiates as thermally activated grain boundary motion takes place. As this occurs, some of the metal remains trapped within pores formed between previous crystallites and therefore is typically an order of magnitude smaller that the final intergranular particulate simply because of the smaller void space that was originally present in between the crystallites (see illustration ①). Some of the former intercrystallite porosity begins to consolidate, forming intragranular pores as grain boundaries sweep beyond the pore structure.

5. Step 5 – the final composite microstructure upon cooling. The spacing of the metallic particles is defined primarily by the sintered grain structure of the ceria, although as mentioned above some of the finer particulate results from being trapped between the voids formed by the former crystallites in the original agglomerate structure.

If our analyses holds true, the microstructure of these composites can be further refined by modifying the size of the original agglomerates in the GNP synthesized powder. As previous work has shown, this can be done by adjusting the glycine-to-nitrate ratio. For example, by reducing this ratio, i.e. running the combustion step in a fuel-lean condition, the average size of the agglomerates and the crystallites that make up these agglomerates can be reduced substantially. This is because combustion takes place at a lower temperature. Conversely, a fuel-rich burn raises the combustion temeperature, yielding powders characterized by large agglomerate and crystallite sizes. Other combustion parameters that can affect combustion temperature will also likely affect the size of the powder in the same way, e.g. the partial pressure of oxygen during combustion (which is typically 0.2 atm since conversion normally takes place in open air). Future studies will attempt to characterize these effects.

A second parameter of interest is the composition of the Ag-CuO phase. Investigations of the wetting characteristics of this material on various ceramic substrates have demonstrated that wetting can be greatly improved at higher concentrations of CuO and/or by employing a secondary wetting agent such as TiO_2 [6]. In a second study, which is currently in progress, we are investigating the effect of Ag-CuO composition on composite microstructure. Ultimately we intend to examine how these microstructures correlate with the inherent fracture toughness properties of the composite.

CONCLUSIONS

The glycine nitrate process was investigated as an in-situ means of preparing $Ag/CuO/CeO_2$ composite powders characterized by a well-dispersed mixture of the consitutent sub-micron phases. The powders were compacted and sintered at a temperature just above the solidus for the

Ag/4vol% CuO alloy employed. As expected, the molten phase enhances the sintering of the ceria matrix. In this way, we were able to develop a uniform, fine-scale dispersion of metallic reinforcement in the final composite microstructure. Specifically, the size, morphology, and spacing of the metal particulate in the densified composite appears to be dependent on the original size and structure of the starting GNP synthesized powders. Additionally work is planned to test this hypothesis and to investigate the effect of Ag-CuO composition on the composite microstructure.

ACKNOWLEDGMENTS

The authors would like to thank Nat Saenz, Shelly Carlson, and Jim Coleman for their assistance. This work was supported by the U. S. Department of Energy, Office of Energy Efficiency and Renewable Energy, FreedomCAR Program. The Pacific Northwest National Laboratory is operated by Battelle Memorial Institute for the United States Department of Energy (U.S. DOE) under Contract DE-AC06-76RLO 1830.

REFERENCES

1. M. Sakai and R.C. Bradt, "Fracture Toughness Testing of Brittle Materials," *International Materials Reviews*, **38** [2], 53-78 (1993).

2. J. Rodel, "Interaction between Crack Deflection and Crack Bridging," *Journal of the European Ceramic Society*, **10** [3], 143-150 (1992).

3. O. Sbaizero and G. Pezzotti, "Tailoring the Microstructure of a Metal-Reinforced Ceramic Matrix Composite," *Transactions of the ASME: Journal of Engineering Materials and Technology*, **122** [3] 363-7 (2000).

4. F. Petit, P. Descamps, M. Poorteman, F. Cambier, and A. Leriche, "Contribution of Crack-Bridging to the Reinforcement of Ceramic-Metal Composites/Definition of an Optimum Particle Size," *Key Engineering Materials*, **206-213** [2] 1189-92 (2002).

5. L.A. Chick, L.R. Pederson, G.D. Maupin, J.L. Bates, and G.J. Exarhos, "Glycine-Nitrate Combustion Synthesis of Oxide Ceramic Powders," *Materials Letters*, **10** [1,2] 6-12 (1990).

6. K.S. Weil, J.S. Hardy, and J.Y. Kim, "A New Technique for Joining Ceramic and Metal Components in High Temperature Electrochemical Devices," *Journal of Advanced Materials*, in press.

7. K.S. Weil and J.S. Hardy, "Use of Combustion Synthesis in Preparing Ceramic-Matrix and Metal-Matrix Composite Powders," *Proceedings of the 28th Annual Conference on Composites, Advanced Ceramics, Materials and Structures*, in press.

8. R.D. Purohit, R.D., B.P. Sharma, K.T. Pillai, and A.K. Tyagi, "Ultrafine Ceria Powders via Glycine-Nitrate Combustion," *Materials Research Bulletin*, **36** [15], 2711-2721 (2001).

9. K.S. Weil, J.Y. Kim, and J.S. Hardy, unpublished research.

ELECTRICAL AND MECHANICAL PROPERTIES OF K, Ca IONIC-CONDUCTIVE SILICON NITRIDE CERAMICS

Yoon-Ho Kim, T. Sekino, T. Kusunose, T. Nakayama, and K. Niihara
The Institute of Scientific and Industrial Research (ISIR),
Osaka University, Mihogaoka 8-1, Ibaraki, Osaka
567-0047, Japan

H. Kawaoka
Energy Electronics Institute, national Institute of
Advanced Industrial Science and Technology (AIST),
Tsukuba Central 2, 1-1-1 Umezono, Tsukuba, Ibaragi
305-8568, Japan

ABSTRACT

The electrical conductivity was providing to structural ceramics by controlling the grain boundary phase. We focused on grain boundary phase of Si_3N_4 ceramics, which can be considered as infinite network for conducting path. The constituents of the conductive path in the sintered samples must be formed as a glassy grain boundary phase during sintering by melting. As the glassy phase, which does not have the particle shape, possesses the radius R_d around zero, it can be expected the critical volume fraction ϕ_c is abruptly reduced compared with particle dispersion method generally used. In this study, we investigated the correlationship of microstructure, sinterability, mechanical properties and ionic conductivity of Si_3N_4 ceramics with alkali alminosilicate grain boundary phase. The result that ionic conductive Si_3N_4 ceramics were fabricated by Pulse Electric Current Sintering (PECS) for controlled composition of grain boundary phase. Fabricated materials by PECS method was indicated very fine microstructure. The Si_3N_4 ceramics had glassy phases between two matrix grains and in multiple grain boundaries, and showed three orders of magnitude higher conductivity than general silicon nitride ceramic.

INTRODUCTION

Alkali silicate glass materials have facility to change the shape at high temperature due to its particular microstructure. This property has made it possible for ionic conductive glass to extent application area. However, this advantage also means that glass materials can't be used at high temperature because it can't keep its shape. Therefore, new technique is required to apply the ionic conductive glass at higher temperature. On the other hand, silicon nitride has been known as one of the representative insulator structural ceramic materials for high-temperature application. However poor sinterability, low electric conductivity and high cost for machining limit its application.[1] To overcome its poor sinterability, oxide additives have been utilized. During sintering, these sintering aids were found to be liquid phase. Also, these oxides exist as glass or crystalline phase in the grain boundary. Although grain boundary phase drops off mechanical properties of Si_3N_4 ceramics at high temperatures. Therefore a many researcher have investigated about grain boundary phase of Si_3N_4 to improve its mechanical properties and thermal properties or oxidation resistance.[2-4]

When conductive particles are dispersed into an insulator matrix, conductivity of the composite changes according to percolation theory in many cases. Near the critical volume, infinite conductive network is formed and the conductivity of the entire composite abruptly increases. The critical volume of dispersed phase is governed by the particle sizes and shapes of primary and second phase. To reduce the critical volume, particle size ratio R_m/R_d is preferable to become large, where R_m and R_d indicate radius of matrix grains and dispersed phase, respectively. Till now, there were many efforts on providing conductivity to insulator ceramics by particle dispersion method. However, second phases with conductivity have been added to these structural insulator ceramic materials to the degree of about 30 vol.% to improve the electric conductivity. These added second phases also more drop off mechanical properties of Si_3N_4 ceramics. However, it has not been found that controlling composition of grain boundary phase provides a new function to silicon nitride ceramics for multi functionality. In this study, we focused on grain boundary of Si_3N_4 materials, which can be considered as infinite network. Some of structural ceramics fabricated with sintering additives have residual grain boundary phase, which consist of glassy or crystalline phase. By controlling the composition of grain boundary phase to form electrical-conductive glassy grain boundary phase, it is certainly expected that the radius of conductive phase can be reduced into near zero, resulting in an abrupt decrease of critical volume, since the glassy phase is not aggregation of particles.

In this study, we investigated the correlationship of microstructure, sinterability, mechanical properties and ionic conductivity of Si_3N_4 ceramics with alkali alminosilicate grain boundary phase. Pulse Electric Current Sintering (PECS) was used for realizing above-mentioned concept. In PECS, densed materials with very fine microstructure can be fabricated because a high-speed heating is expected.

EXPERIMENTAL PROCEDURE

Processing and Sample Preparation

α-Si_3N_4 (SN-E10, Ube Industries Co., Ltd., Ube, Japan) with high purity was used as matrices.

α-Al$_2$O$_3$ (99.99% pure, Taimicron TM-DAR, Taimei Chemicals Co. Ltd., Nagoya, Japan), K$_2$CO$_3$, CaCO$_3$, and SiO$_2$ (99% pure, KOJUNDO Chemical Laboratory, Saitama, Japan) were used as sintering additives. These components were weight into a composition 90 Si$_3$N$_4$ / 10 (40 XO⁻ 30 Al$_2$O$_3$⁻ 30 SiO$_2$ [mol%], X: K$_2$, Ca) [wt%]. The sample names designedly the matrix and additives such as SNK, SNCa. They were mixed by wet ball-milling in a polyethylene pot with ethanol for 24 hours. Then, the mixed powders were dried and crushed by dry ball-milling in the polyethylene pot for 12hours to eliminate the agglomerates. These powders were packed into carbon mold with 20 mm in inside-diameter and sintered by a Pulsed Electric Current Sintering (PECS) (Dr. SINTER^{TM}, SPS-2040, IZUMI TECHNOLOGY Co., Ltd., Hokkaido, Japan). The Si$_3$N$_4$ ceramics were sintered at 1600 °C for 5 minutes under an applied pressure of 50 MPa in N$_2$ atmosphere.

Physical and Mechanical Property Measurement

Sintered materials were cut and ground into 2×3×20 mm rectangular specimens and polished with diamond pastes. The density of sample was measured by Archimedes method using toluene. Crystalline phases of sintered specimens were determined by XRD analysis using a RIGAKU Rotaflex diffractometer (RU-200B, RIGAKU Co., Ltd., Japan) with a CuKα radiation (l=0.15481 nm) operated at 50 kV and 150 mA. Vickers indenter (AVK-C2, Akashi CO., Ltd., Tokyo, Japan) was employed to obtain hardness and fracture toughness of sintered samples. Indentation load was 196 N and duration time was 15 seconds. The value of fracture toughness was obtained by Indentation Fracture (IF) method. The average values of hardness and fracture toughness were obtained from 10 measurements. Flexural strength was measured by three point bending test. The edges of the tensile side of the bend bars were chamfered at 45° by send paper, and then three point bending test was performed by a universal testing machine (AUTOGRAPH AG-10TC, Shimadzu Co., Ltd., Kyoto, Japan). Span was 15 mm and crosshead speed 0.5 mm/min. All data points are the average of at least 6 measurements. Such samples were coated with gold prior to SEM (S-500, Hitachi Co., Ltd., Japan) observation.

Electrical Property Measurement

The electrical conductivity was calculated from the resistance measured by two-terminal A.C. method between 100 and 1000 °C at frequency of 10 KHz in air atmosphere by using LCR meter (HP-4284A, Hewlett Packard Co., USA). As-sintered samples (20 mm in diameter) were ground by diamond wheels and polished with diamond slurry to less than 1 mm in thickness. Before measuring, platinum electrodes were pasted at the center of both side of specimens with 15 mm diameter and heat-treated at 700 °C for 1 minute.

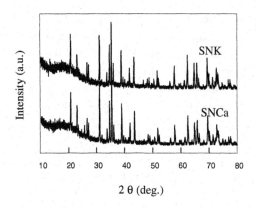

Fig. 1. XRD data for Si_3N_4 ceramics with $XO-Al_2O_3-SiO_2$ (X: K_2, Ca)

RESULT AND DISCUSSION

Materials are successfully fabricated by PECS for controlled composition of grain boundary phase Si_3N_4 ceramics. From XRD result for Si_3N_4 ceramics, only α-and β-Si_3N_4 was identified and no crystalline phase in grain boundary as shown in Fig. 1. The broad bulge in the lower 2θ angle area obviously indicates the presence of glassy phase. It can be thought that the sintering additives formed glassy phase in the grain boundary due to the fast cooling of the PECS process. α-and β-Si_3N_4 crystalline phases were calculated by X-ray diffraction analysis of the sintered materials as shown in Table I.

Table I. Physical and mechanical properties of each sample

Sample Name	$\alpha : \beta$	Density (%)	Young's Modulus (GPa)	Hardness (GPa)	Fracture Toughness (MPa·m$^{1/2}$)	Flexural Strength (MPa)
SNK	91: 9	77.0	276	13.7	3.56	631
SNCa	88:12	99.2	319	15.3	4.59	706

Physical and mechanical properties of each sample are summarized in Table I. The relative density of SNCa is above 99 %, however SNK indicates low relative density below 80 %. The SNCa is indicated high young's modulus above 319.6 GPa such as Table I. However, the SNK shows very low young's modulus about 276 GPa because of low relative density. The flexural strength is also shown in Table I. The SNK and the SNCa are shown lower value compare to conventional silicon nitride, because those materials are contained low β-Si_3N_4 phase. In the case of Si_3N_4 ceramic added CaO, the reason of lower flexural strength is thought that the grain boundary phase is more week glass because the silica network is disrupted by adding Ca. Generally CaO is known to be a potent silica network modifier, causing a decrease in glass viscosity by several orders of magnitude for CaO additions less than 1.0 mol%.[5] Generally the hardness of Si_3N_4 ceramics increases with the reducing grain size, and thus the α-Si_3N_4 ceramics

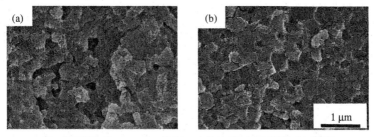

Fig. 2 SEM micrographs of fractured surface (a) SNK (b) SNCa

showed higher hardness than the β-Si$_3$N$_4$ ceramics.[6,7] Greskovoch et al. reported that the dense Si$_3$N$_4$ ceramic with 15 % β-phases showed about 1.3 times higher Vickers microhardness than 100 % β-specimen.[8] The SNCa indicates high value of hardness about 15.3 GPa. The SNK is shown low value compare to the conventional silicon nitride because of low relative density. The fracture toughness is shown in Table I. Effect of grain morphology of β-phases on fracture toughness has been extensively investigated.[9-14] Generally, fracture toughness increases β-Si$_3$N$_4$ content increases.[9,10] The SNK and the SNCa indicate low value because those materials are contained low β-Si$_3$N$_4$ phase. However, the possibility of increasing mechanical properties is very high because it can be thought that amount of β-Si$_3$N$_4$ phase can controlled as sintering conditions.

Fig. 2 shows the SEM photographs of fractured surfaces after three point bending test. All as-sintered Si$_3$N$_4$ specimens with XO-Al$_2$O$_3$-SiO$_2$ possessed very fine microstructures due to the very short sintering time of the PECS method. The SNK and the SNCa are almost α-Si$_3$N$_4$ ceramic and lots of pores were observed in the SNK, which well agree with the result of the XRD analysis and relative density.

The variation of electrical conductivity of the SNK, SNCa and Si₃N₄ ceramics sintered with 3 mol% Al₂O₃ and 6 mol% Y₂O₃ additives were shown in Fig. 3. The electrical conductivity of SNK

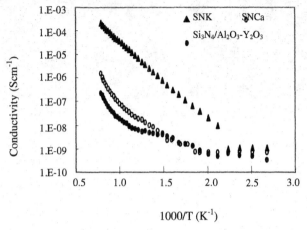

Fig. 3 Electrical Conductivities of SNK, SNCa and Si₃N₄/Al₂O₃-Y₂O₃ ceramics between 100 and 1000 °C

ceramic had three orders of magnitude higher than that of Si₃N₄/Y₂O₃-Al₂O₃ ceramics at 1000 °C. The Arrhenius plot of the conductivity for SNK ceramics had no flexural point, which was shown in the variation of conductivity for Si₃N₄/Y₂O₃-Al₂O₃ ceramics. It can be thought that the potassium cation (K⁺) move through the glass network in grain boundary, which is looser compared to crystalline, and work as conductive carrier. In the case of Si₃N₄ added Ca²⁺ ion had one order of magnitude higher than that of conventional Si₃N₄ ceramics at 1000 °C. SNCa ceramics had many flexural points. The Arrhenius plot of the conductivity for SNCa ceramics is similar to Si₃N₄/Y₂O₃-Al₂O₃ ceramics. The conduction mode is great different between SNK and Si₃N₄ ceramics with Al₂O₃-Y₂O₃ additives implies such as Fig. 3. This change of activation energy between the higher and lower temperature rages in Si₃N₄/Al₂O₃-Y₂O₃ had already been reported for Si₃N₄ ceramics with MgO or Al₂O₃-Y₂O₃ additives[15,16] and for SIALON ceramics.[17]

CONCLUSIONS
The Si₃N₄ ceramic with K⁺, and Ca²⁺ ionic conductivity was fabricated by controlling the composition of grain boundary phase. Fabricated materials by PECS method is indicated very fine microstructure. The Si₃N₄ ceramics had glassy phases between two matrix grains and in multiple grain boundaries, and showed one and three orders of magnitude higher conductivity than

Si_3N_4/Al_2O_3-Y_2O_3 ceramics at 1000 °C, respectively. The advantage of this concept is a facility of the processing and that much lower addition is required to express the electric conductivity than the particle dispersion method. Moreover, this processing is very simple, where the providing conductivity was accomplished by only changing or adding the sintering additives. As the multi-functionalizing by this concept indicates certain possibility can be applied to a number of matrixes require sintering additives.

REFERENCES

[1]R. Kossowsky, D. C. Miller, and E. S. Diaz, "Tensile and Creep Strength of Hot Pressed Si_3N_4," *J. Mater. Sci.*, **10** 983-997 (1975).

[2]H. Schubert, J. Deuerlein, U. Neidhardt, H. Lange, F. Aldinger, and E. Muschelknautz, "Effect of Powder Modification on Sinterability, Microstructure and Properties of Si_3N_4," *Key Eng. Mater.*, **89-91** 141-146 (1994).

[3]V. Vandeneede, A. Leriche, F. Cambier, H. Pickup, and R. J. Brook, "Sinterability of Silicon Nitride Powders and Characterisation of Sintered Materials," *Non-Oxide Tech. and Eng. Ceram., Proc. of the International Conference*, 53-68 (1986).

[4]K. Strecker, R. Gonzaga, and M. Hoffmann, "Sintering of Silicon Nitride Ceramics with Additive Mixtures Based on Yttria, Aluminum Nitride and Alumina," *Key Eng. Mater.*, **189-191** 110-119 (2001).

[5]H. J. Kleebe, "Structure and Chemistry of Interfaces in Si_3N_4 Ceramics Studied by Transmission Electron Microscopy," *J. Ceram. Soc. Jpn.*, **105** 453 (1997).

[6]R. F. Coe, R. J. Lumby, and M. F. Pawson, "Special ceramics 5," *edited by P. Popper*, 361 (1972).

[7]A. K. Mukhopadhyay, S. K. Datta, and D. Chakraborty, "Hardness of Silicon nitride and Sialon," *Ceram. Int.*, **17** 121-128 (1991).

[8]C. Greskovich and G. E. Gazza, "Hardness of dense α- and β-Si_3N_4 ceramics," *J. Mater. Sci. Lett.*, **4** 195 (1985).

[9]M. Mitomo, N. Hirosaki, and H. Hirotsuru, "Microstructural design and control of silicon nitride ceramics," *MRS Bull.*, Feb. 38-41 (1995).

[10]F. F. Lange, "Relation between strength, fracture energy, and microstructure of hot-pressed Si_3N_4," *J. Am. Ceram. Soc.*, **56** 518 (1973).

[11]K. T. Faber and A. G. Evans, "Crack deflection processes-□Theory," *Acta Metall.*, **31** 565-576 (1983).

[12]K. T. Faber and A. G. Evans, "Crack deflection processes-□Experiment," *Acta Metall.*, **31** 577-584 (1983).

[13]E. Tani, S. Umebayashi, K. Kishi, Kazuo kobayashi, and M. Nishijima, "Gas-pressure sintering of Si_3N_4 with concurrent addition of Al_2O_3 and 5 wt% rare earth oxide: High fracture

toughness Si_3N_4 with fiber-like structure," *Am. Ceram. Soc. Bull.*, **65** 1311 (1986).

[14]P. F. Becher, S. L. Hwang, and C. H. Hsueh, "Using microstructure to attack the brittle nature of silicon nitride ceramics," *MRS Bull.*, Feb. 23 (1995).

[15]Y. Katano, H. Ohno, and H. Katsuta, "Electrical Properties and Microstructures of Hot-Pressed Silicon Nitride," *J. Nucl. Mater.*, **141-143** 396-400 (1984).

[16]H.Ohno, T. Nagasaki, Y. Katano, J. Tateno, and H. Katsuta, "Electrical properties and phase stabilities of some ceramics irradiated by neutrons and ions," *J. Nucl. Mater.*, **155-157** 372-377 (1988).

[17]G. R. Rao, S. A. Kokhtev, and R. E. Leohman, "Electrical and thermal conductivity of sialon ceramics," *Am. Ceram. Soc. Bull.*, **57** 591-595 (1978).

ELECTRICAL RESISTANCE MEASUREMENTS OF CONDUCTIVE OXIDE DISPERSED GLASS COMPOSITES FOR SELF DIAGNOSIS

Byung-Koog Jang and Hideaki Matsubara
Japan Fine Ceramics Center
2-4-1, Mutsuno, Atsuta-ku, Nagoya, 456-8587, JAPAN

ABSTRACT
The conductive composites consisting of ceramic fiber/RuO_2/glass matrix were fabricated by sintering at 850°C for use of self-diagnosis materials. Three types of ceramic fibers were added to the conductive composites for reinforcement. The sensing properties of self-diagnosis materials were investigated in real time during tensile testing by measuring the electrical resistance. It is shown that the excellent sensing ability based on electrical resistance changes in the low strain range was due to brittle fracture of the glass matrix. The change of electrical resistance depends strongly on the volume percent of fiber reinforcement.

INTRODUCTION
Recently, many studies regarding in-situ detection of electrical resistance changes using conductive composites have been reported. [1-7] These conductive composites have usually been developed for use as self-diagnosis materials in civil concrete structures by in-situ detection of electrical resistance. [8-9] Generally-speaking, conductive composites for self-diagnosis consist of a combination of a conductive material, a non-conductive matrix and an insulator for surface coating.

If these self-diagnosis materials are subjected to loading or damage, their electrical resistance changes. This change in conductive composites is due to deformation or changes in the connectivity between the conductive particles that provide conduction paths for the electrical current. Consequently, measurement of electrical resistance change provides self-diagnosis sensing ability to the self- diagnosis materials, in a manner similar to other sensing methods such as optical fibers and strain gages. If the self-diagnosis materials experience the mechanical damage due to a tensile stress, the amount of electrical resistance change induced in the self-diagnosis materials can be used to assess the degree of damage. [10-13] For self-diagnosis materials, carbon fibers or carbon particles are often used as the electrically conducting phase and epoxy as the matrix. [14-18]

However, the sensing abilities of CFRP (carbon fiber reinforced plastics) composites or carbon particles-dispersed composites based on electrical resistance change is insufficient, especially at low strains (< 0.5%), because the epoxy matrix is non-brittle. For this reason, the purpose of the present work is to develop a new type of conductive composites displaying significant change in electrical resistance at low strains < 0.5%.

To accomplish this purpose, conductive oxide dispersed brittle composites were designed by combining ceramic fibers (as forming reinforcement), RuO_2 particles (as conductive oxide), glass (as the brittle matrix) and an outer insulating layer of FRP as shown Fig. 1(a). We investigated the influence of fiber content and fiber type on electrical resistance changes in these composites during tensile loading.

EXPERIMENTAL

The conductive materials used were the pastes made from fine RuO_2 particles and glass powders dispersed in an organic solvent. Glasses of the $PbO-SiO_2-B_2O_3-Al_2O_3$ system having low softening temperatures were used to prevent the degradation of the reinforcing ceramic fibers during sintering. The chemical composition of the conductive paste used, which has a resistance of 40 Ω, is given in Table 1. Table 2 lists the properties of fiber types used for reinforcement of the glass matrix.

Table I. Chemical composition of the conductive glass

Chemical composition (wt%)				
PbO	SiO_2	B_2O_3	Al_2O_3	RuO_2
30	20	6	1	43

Table II. Properties of reinforced ceramic fibers

Maker	Daimei	Asahi Glass
Grade	HT	E-glass
Fiber	alumina	glass fiber
Tensile modulus (GPa)	220	72.5
Tensile strength (GPa)	2.5	3.4
Elongation (%)	1.1	4.8

The fabrication procedure of the fiber reinforced conductive composites was as follows. The continuous ceramic fibers were dipped into the conductive RuO_2/glass paste. Green bodies containing 3~14 vol% fibers embedded into the RuO_2/glass matrix were formed into cylindrical rods. After drying at 130°C for 2 h, the conductive green bodies were sintered at 850°C for 30 min. Electrical copper wires were connected at both ends of the sintered conductive bodies and silver paste was used to ensure good electrical contact. To provide insulation and protect against any handling damage, glass fiber reinforced epoxy resins were coated on the surfaces of the sintered specimens, and dried at 130°C. Afterward, ring-type steel tabs were applied to both ends of the specimen for clamping during tensile tests. The ring tabs were filled with expansive cement to bond the tabs and specimen together solidly. The samples for tensile testing had a diameter of 2 mm and a length of 200 mm as shown in Fig. 1(b).

insulating layer

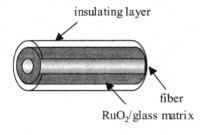 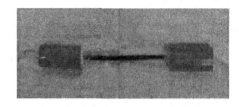

fiber

RuO_2/glass matrix

(a) schematic of composites structure   (b)   tensile specimen

Fig. 1. Conductive oxide dispersed composites.

Fig.2. Setup view for tensile testing.

The electrical resistance was measured by a two-terminal method using an electrical resistance analyzer. An extensometer was attached to the center of the specimen to measure strain as shown in Fig. 2. The strain and electrical resistance of the specimens were measured simultaneously with increasing tensile load. The microstructures of the specimens were observed by SEM and the elemental composition was analyzed by EDS.

RESULTS AND DISCUSSION

Figure 3 shows a typical microstructure of the polished surface of a 6vol% Al_2O_3 fiber reinforced RuO_2/glass conductive composite. The vertical dark regions are the embedded Al_2O_3 fibers, whereas the surrounding regions are the RuO_2 containing glass matrix. The fine RuO_2 particles dispersed throughout the glass matrix can support electrical conduction via points of contact between them. Figure 4 (a) shows the microstructure of the cross-sectional surface of the 6vol% Al_2O_3 fiber reinforced RuO_2/glass conductive composite. The dark circle is the embedded Al_2O_3 fiber, and the lighter area is the RuO_2/glass matrix. Figure 4 (b) shows the results of quantitative analysis of the constituent elements, namely Al, Si, Ru and Pb for the fiber, interface and matrix in Fig. 4 (a). No reaction between the Al_2O_3 fibers and glass matrix could be detected, so that it can be concluded that there is good bonding between the Al_2O_3 fiber and matrix. Figure 5 shows the influence of the amount of Al_2O_3 fiber in Al_2O_3 fiber/RuO_2/glass composites on the electrical resistance change at low strains up to 0.5% during tensile loading. The electrical resistance change is defined as $\Delta R/Ro$, where ΔR and Ro are the electrical resistance increase and initial electrical reesistance, respectively. The load behavior in response for increasing strain is a non-linear and exhibits elastic-plastic deformation characteristics because of the reinforcement effect of the embedded fiber. The stress curve is linear at strains of less than about 0.1%, that is, the elastic region. More importantly, the stress increases in a stepwise manner due to the fiber

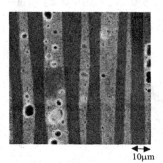

Fig. 3. SEM micrograph of polished surface of Al_2O_3 fiber/RuO_2/glass composites.

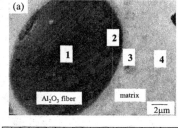

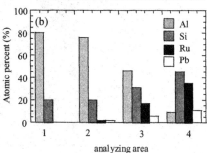

Fig. 4. Elemental mapping of fiber and matrix by EDS for Al_2O_3 fiber/RuO_2/glass composites; (a) SEM of cross-section and (b) quantitative analysis of elements in (a).

reinforcement at strains > 0.1% in Fig 5. (c) and Fig 5. (d).

The electrical resistance overall increased with increasing strain in accordance with the applied loading. The increase in the electrical resistance with increasing loading indicates fracture of or damage to the specimen has occurred. Fracture of the 3 vol% Al_2O_3 fiber reinforced sample in Fig 5 (a) occurred in the low strain range, i.e., < 0.4 % and became insulating at 0.3% strain. The magnitude of the increase in the electrical resistance change with increasing amount of Al_2O_3 fibers is relatively small.

The reason for the steady increase in resistance with increasing load is the gradual small occurrence of brittle fracture in the matrix due to the strengthening effect of Al_2O_3 fibers, which results in a small amount of deformation or changes in the conduction paths between RuO_2 particles.

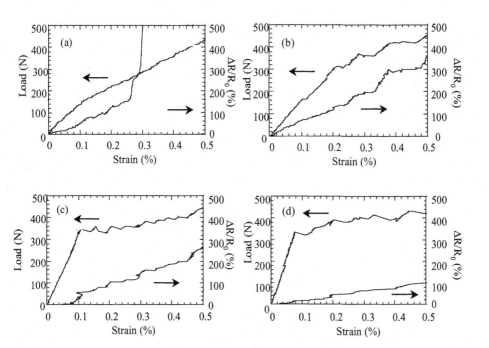

Fig. 5.   Influence of amount of Al_2O_3 fibers on electrical resistance change and load as a function of strain during tensile tests for Al_2O_3fiber/RuO_2/glass composites; (a) 3 vol% (b) 6 vol% (c) 10 vol% and (d) 14 vol% of Al_2O_3 fibers.

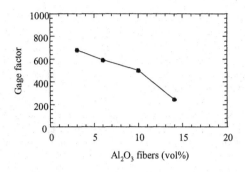

Fig. 6. Comparison of gage factor at 0.2 % strain as a function of volume of Al$_2$O$_3$ fibers for Al$_2$O$_3$ fiber/RuO$_2$/glass composites.

The results of Fig. 5 show that the electrical resistance change and load curve in response to the strain increase depend strongly on the amount of Al$_2$O$_3$ fibers. The values of electrical resistance change in the present specimen are hundreds of times greater than those of CFRP or carbon particle reinforced epoxy composites[16-18], resulting in a high sensitivity in the low strain region for our now composites. This remarkable increase in electrical resistance is caused by fracturing of the brittle glass and the associated decrease in the number of conductive paths between adjacent RuO$_2$ particles.

Figure 6 shows a plot of gage factors at 0.2 % strain as a function of the amount of alumina fibers for Al$_2$O$_3$ fiber/RuO$_2$/glass composites. The gage factor is defined as ($\Delta R/R_0$) / ε, where $\Delta R/R_0$ and ε are the electrical resistance change and the strain, respectively, and is dimensionless. The gage factor usually serves as an indicator of the sensitivity of a self-diagnosis material. High gage factors mean that the electrical resistance change is large, corresponding to high sensitivity. The gage factor of the present specimens decreased with increasing the amount of Al$_2$O$_3$ fibers for Al$_2$O$_3$ fiber/RuO$_2$/glass composite.

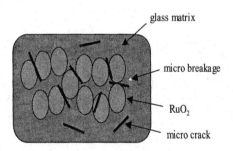

Fig. 7. Mechanism of electrical resistance change by generation of microcracks in the matrix for RuO$_2$/glass conductive composites under applied tensile loading.

The mechanism for the change in electrical resistance can be illustrated as shown in Fig. 7. If some damage or tensile loading is applied to a conductive specimen, microbreakages and/or microdeformation of conductive RuO_2 particles occurred frequently due to brittle fracture at the glass matrix/particles interface. Consequently, these microchanges in the conduction paths between RuO_2 particles causes the electrical resistance to increase, resulting in large increase in electrical resistance. An increase in the applied load or damage results directly in an increase of the electrical resistance change. The above results provide encouraging evidence that conductive oxides dispersed in suitable composites can be tailored for use as self-diagnosis materials. The damage levels in such materials can be easily monitored by monitoring the changes in their electrical resistance.

CONCLUSIONS

Al_2O_3 or glass fiber reinforced RuO_2/glass composites were fabricated by sintering at 850 °C with the aim of using them as self-diagnosis materials. The electrical resistance changes of specimens increased remarkably with increasing strain in the low strain range, < 0.5%, during tensile loading. The high variation in electrical resistance was caused by increasing number of microcracks in the brittle glass matrix and the consequent decrease in the number of conductive paths between adjacent RuO_2 particles. The electrical resistances of the conductive specimens are largely dependent on the volume percent of reinforcing fibers. The magnitudes of the electrical resistance change and gage factor for higher volume Al_2O_3 fiber reinforced specimens were low than those of lower volume Al_2O_3 fiber reinforced specimens.

ACKNOWLEDGMENTS

The authors would like to acknowledge the financial support of the New Energy and Industrial Technology Development Organization (NEDO) of the Japanese government.

REFERENCES

[1]T.Prasse, F.Michel, G.Mook, K.Schulte and W. Bauhofer, "A comparative Investigation of Electrical Resistance and Acoustic Emission during Cyclic Loading of CFRP Laminates," *Composites Science and Technology*, **61**, 831-35 (2001).

[2]K. Schulte, "Load and Failure Analysis of CFRP Laminates by Means of Electrical Resistivity Measurements," *Composites Science and Technology*, **36**, 63-76 (1989).

[3]K.Schulte and C. Baron, "Load and Failure Analysis of CFRP Laminates by Means of Electrical Resistivity Measurement," *Composites Science and Technology*, **36**, 63-76 (1989).

[4]X.Wang and D.D.L. Chung, "Sensing Delamination in a Carbon Fiber Polymer-Matrix Composite During Fatigue by Electrical Resistance Measurement," *Polymer Composite*, **18** [6] 692-700 (1997).

[5]X.Wang and D.D.L. Chung, "Fiber breakage in Polymer-Matrix Composite During Static and Fatigue Loading, Obserbed by Elecrical Resistance Measurement," *Journal of Material Research*, **14** [11] 4224-229 (1998).

[6]A. Todoroki, Y. Tanaka and Y. Shimamura, "Electric Resistance Change Method for Identification of Embedded Delamination of CFRP Plates," *Journal of Material Science Japan*, **50** [51] 495-501 (2001).

[7]J.B.Park, T. Okabe, A. Yoshimura, N. Takeda and W. A. Curtin, "Quantitative evaluation of the electrically Conductive Internal Network in CFRP Composites"; pp. 40-50 in Proc. of SPIE

Vol 4704, Edited by A.R. Wilson and H.Asanuma, 2002.

[8]M.Sugita, H. Yanagida and N. Muto, "Materials Design for Self-Diagnosis of Fracture in CFRP Composite Reinforcement," *Smart Materials and Structures*, **4**, A52-57 (1995).

[9]D.D.L. Chung, "Self-monitoring structural materials," *Materials Science Engineering*, **R22**, 57-78 (1998).

[10]S.Wang and D.D.L. Chung, "Mechanical Damage in Carbon Fiber Polymer-Matrix Composite, studied by Electrical Resistance Measurement," *Composite Interface*, **9** [1] 51-60 (2002).

[11]P.E. Irving and C. Thiagarajan, "Fatigue Damage Characterization in Carbon Fibre Composite Materials Using an Electrical Potential Technique," *Smart Materials and Structures*, **7** [6] 456-66 (1998).

[12]H. Wittich, K. Schulte, M. Kupke, H. Kliem and W. Bauhofer, "The Measurement of Electrical Properties of CFRP for Damage Detection and Strain Recording"; pp. 447-57 in Proc. 2nd ECCMCTS, Hamburg, 1994.

[13]O. Ceysson, M. Salvia and L. Vincent, "Damage Mechanism Characterization of Carbon Fibre/Epoxy Laminates by Both Electrical Resistance Measurements and Acoustic Emission Analysis," *Scripta Material*, 34, 1273-80 (1996).

[14]N. Muto, H. Yanagida, M. Miyayama, T. Nakatsuji, M. Sugita and Y. Ohtsuka, "Foreseeing of Fracture in CFGFRP Composites by the Measurement of Residual Change in Electrical Resistance," *Journal of the Ceramic Society of Japan*, **100** [4] 585-88 (1992).

[15]X.Wang and D.D.L. Chung, "Continuous Carbon Fibre Epoxy-Matrix Composite as a Sensor of its Own Strain," *Smart Materials and Structures*, **5** [6] 796-800 (1996).

[16]Y.Okuhara, S.G. Shin, H. Matsubara and H. Yanagida, "Self-Diagnosis of the Composite Containing Electrically Conductive Phase," *Transactions of the Materials Research Society of Japan*, **25** [2] 581-84 (2000).

[17]Y.Okuhara, S.G. Shin, H. Matsubara, H. Yanagida and N. Takeda, "Self Diagnosis function reinforced composite with Conductive Particles"; pp. 44-51 in Proc. of SPIE Vol 4234, Edited by A.L. Gyekenyesi, S.M. Shepard, D.R. Huston, A.E. Aktan and P.J.Shull, 2002.

[18]S.G. Shin, H. Matsubara, Y.Okuhara, H. Yanagida and N. Takeda, "Self-Monitoring of FRP Using Electrical Conductivity of Carbon Phase"; pp. 995-98 in Proc. 6th Japan International SAMPE Symposium, Tokyo, 1999.

Geopolymers

A CONCEPTUAL MODEL FOR SOLID-GEL TRANSFORMATIONS IN PARTIALLY REACTED GEOPOLYMERIC SYSTEMS

John L. Provis, Jannie S. J. van Deventer* and Grant C. Lukey.
Department of Chemical and Biomolecular Engineering
The University of Melbourne
Victoria 3010
AUSTRALIA.

ABSTRACT

Recent research on the microstructure of geopolymers has focused on using fully reacted matrices as a reference for investigating the effect of different processing conditions. Nevertheless, all practical geopolymer systems are partially reacted with residual aluminosilicate particles being embedded in a matrix of hardened geopolymeric gel. Much work has been focused on understanding the microstructure of this gel phase, but unfortunately little work has been devoted to the interface between the reacted and unreacted phases and the associated transport processes. This paper reviews these transport and reaction mechanisms, proposes a simplified mathematical model for selected reactions, and presents a hypothesis for the nanostructure of the formed material. From existing published experimental data and by comparison with related hydrothermal mineral synthesis systems, it may be concluded that a significant component of the binder phase formed in geopolymerization is nanometre-sized crystalline structures resembling the nuclei around which zeolites crystallize. Agglomeration of these nanocrystallites with other less-ordered aluminosilicate structures results in the formation of a high-performance mineral binder or geopolymer. The use of alkaline silicate solution as an activator gives a product with higher strength than if alkali hydroxides are used, and this is attributable to the rapid nucleation of solid products immediately surrounding the dissolving aluminosilicate source particles. These form a matrix binding the geopolymeric product together, and reduce the likelihood of the meso- to macroscopic flaws responsible for the low strength of some geopolymers. The identification of zeolitic nanocrystals within the geopolymeric binder is potentially a highly significant observation, as it provides a link between the chemical composition and engineering properties of geopolymeric materials. Mechanical strength testing is not an ideal means of determining the success or otherwise of geopolymerization on a nanostructural level, but if carried out correctly is capable of providing data of some value in structural analyses.

INTRODUCTION

'Geopolymer' is the name that since the late 1970s has been applied to a wide range of alkaline- or alkali-silicate-activated aluminosilicate binders of composition $M_2O.mAl_2O_3.nSiO_2$, usually with $m \approx 1$ and $2 \leq n \leq 6$, and where M represents one or more alkali metals [1]. Some geopolymers also contain alkaline earth cations in addition to the alkali metal cations, although some controversy remains regarding the specific role of these cations within the structure as will be discussed later in this paper. Geopolymers are formed by alkali activation of aluminosilicates obtained from industrial wastes, calcined clays, melt-quenched glasses, natural minerals, or mixtures of materials from two or more of these categories. Filler materials including conventional concrete aggregates such as basalt may be used to enhance desired properties including strength and density, or to minimize shrinkage on drying. Activation is achieved by mixing solid aluminosilicate materials with alkali metal hydroxide or silicate solutions at ambient or elevated temperature.

The geopolymeric binder phase is generally described as 'X-ray amorphous.' Many authors have noted formation of phases described as either semicrystalline or polycrystalline, particularly in products synthesized at a higher temperature. However, the chemical and physical nature of these phases has rarely been subjected to detailed analysis, and is very difficult to determine due to the complex and intergrown nature of the binder phases and the presence of significant quantities of unreacted raw materials. Also, the historically application-driven nature of most work in this field has meant that most workers have focused primarily on the engineering properties of geopolymers rather than on detailed analysis of their nanometre-level structure. This has led to the development of a significant body of empirical data described by correlations between physical properties and synthesis parameters but with limited fundamental insight into the mechanisms by which these correlations occur. However, it is only by obtaining such an insight that the means of development of desired physical properties during the setting process may be understood, thus providing the motivation for the current work.

Hydrothermal techniques have been used in mineral synthesis processes for in excess of 50 years, particularly in the production of a wide range of synthetic aluminosilicate structures both with and without naturally-occurring analogues. Some of these synthetic compounds are zeolitic, and have therefore received much attention over the past several decades in their applications as catalysts and molecular sieves. Much is therefore known about the hydrothermal synthesis of zeolites, and this knowledge may be used to gain a further understanding of the chemistry of related systems. It has been stated that geopolymers may be viewed as the amorphous analogue of zeolites [2], as synthesis may be carried out under similar hydrothermal conditions and the presence of 'zeolitic water' has been noted in DTA experiments [3]. Early in the development of geopolymers as a commercial product, synthesis temperatures of up to 150°C were used, making geopolymerization a true hydrothermal process. More recently, ambient temperature synthesis has been shown in some circumstances to give a stronger product, but the classification of geopolymerization as a hydrothermal process remains valid at these lower temperatures.

Taking into account the existing confusion and disagreement regarding the exact chemical nature of geopolymers which is currently hindering the commercialization and application of this highly promising technology, the purpose of this paper is therefore threefold: To propose a new structural model for geopolymers that is consistent with existing knowledge in the wider field of aluminosilicate chemistry as well as observed physical and engineering properties of geopolymers; to re-examine existing experimental results in the light of this new structural theory; and to motivate and develop the first ever mathematical model for the process of geopolymerization in light of this new structural model.

CLASSIFICATION OF GEOPOLYMERS

The geopolymeric binder phase is often categorized as an aluminosilicate gel, and due to compositional similarities it has therefore often been proposed that this structure is related to the aluminosilicate precursor gels from which zeolites are hydrothermally generated. Due probably to the difficulties inherent in detailed structural analysis of gel-phase systems, this suggestion has never been subjected to rigorous investigation. The fact that zeolitic materials are often detected in geopolymeric systems suggests that this proposal is definitely worthy of further discussion.

Such discussion will form a central part of this paper, and the key conclusions of this work will be based heavily on understanding drawn from zeolite chemistry.

Some authors have also described the geopolymer phase as 'glassy.' This claim has never been fully theoretically justified, and seems to be based entirely on the apparent amorphicity of the geopolymeric binder. However, the fundamental difference between geopolymers and glasses lies in the method of hardening – geopolymerization is a solution-mediated chemical reaction process in which solidification occurs by a series of precipitation reactions from a highly supersaturated solution, while glass formation from aluminosilicate melts is a physical cooling to solidify a liquid. With the physicochemical nature of the starting materials differing so widely, the products formed will also be expected to display fundamental differences in chemical and physical structure. Swier et al. [4] observed that geopolymeric structure development as measured by changes in specific heat capacity continues for at least 50 hours after setting ("vitrification"), and Rahier et al. [5] stated that the rate of reaction "seems not to be influenced by the vitrification." These observations present another fundamental difference between hydrothermal geopolymerization and melt-quenching, in which further structure development is greatly impeded by solidification.

The other standard category of materials into which geopolymers are increasingly being placed is 'ceramic,' as evidenced by the increasing popularity of American Ceramic Society events as forums in which Symposia on geopolymers are held. The low temperatures at which geopolymeric setting takes place provides an obvious difference from the traditional definition of ceramics as materials that are formed through a firing process, as does the fact that the setting is a chemically-induced rather than a physically-induced process. Nevertheless, as the field of ceramics expands its borders to include a much wider range of advanced and composite materials, the brittle nature of geopolymers and the fact that they are often synthesized from aluminosilicate clays will inevitably lead to comparisons with ceramic materials. There may be much to learn regarding measurement, modification and optimization of the engineering properties of geopolymers from comparisons with traditional ceramics, particularly with regard to the use of geopolymeric binders as the basis for high-performance composite materials. However, as this paper is primarily focused on description of structure by analysis of the synthesis mechanisms of geopolymers, comparisons in the current investigation will generally be made with other hydrothermally synthesized materials such as zeolites and their precursor gels rather than traditional ceramics.

XRD ANALYSIS OF 'AMORPHOUS' ALUMINOSILICATES

The primary value of powder X-ray diffraction (XRD) analysis of materials is that the angles at which diffraction of a given wavelength of radiation occurs from a polycrystalline sample are characteristic of particular interlayer spacings in the crystal structure of the sample, according to Bragg's Law. Therefore, materials with similar crystal structures and thus similar interlayer spacings will produce similar diffraction patterns. Smaller crystallites give broader diffraction peaks, with the diffractogram of a fully amorphous material displaying very broad, featureless peaks, or 'humps.' However, the exact boundary between 'crystalline' and 'amorphous' materials is very difficult to determine. The International Union of Crystallography defines a crystal as "any solid that gives a discrete X-ray diffraction diagram" [6]. This means that a crystal is defined by its measurable properties. The definition is clearly then dependent on the measurement technique used because X-ray amorphous materials can display crystallinity readily observable via electron diffraction. The importance of the vagueness of this definition in the context of characterization of geopolymeric systems is that even 'X-ray amorphous' structures will produce a diffractogram that is to some extent characteristic of the particular structure present. These diffractograms are unlikely in themselves to provide positive identification of an unknown material as is the case with fully crystalline samples. However, comparison of the diffractogram of a sample of unknown structure with those of materials of similar chemical composition and known structure can at least show with which materials the unknown sample does or does not share significant structural features.

The major feature of XRD powder diffraction patterns of geopolymers is a largely featureless 'hump' centred at approximately 27-29° 2θ. An example of a typical series of geopolymer X-ray diffractograms is presented in Figure 1. Numerous other examples may be found in the literature. However, the most notable feature of all diffractograms of geopolymers is that regardless of the solid aluminosilicate source (metakaolin with or without added calcium, fly ash or blast furnace slag), activating solution (sodium or potassium hydroxide at different concentrations, with or without soluble silicate), and curing conditions (time, temperature and relative humidity) used, the broad hump centred at around 27-29° 2θ is present in every case. This must therefore be considered the distinguishing feature of the diffractogram of any geopolymer, and so its identification becomes central to the potential use of XRD in determination of geopolymeric microstructure.

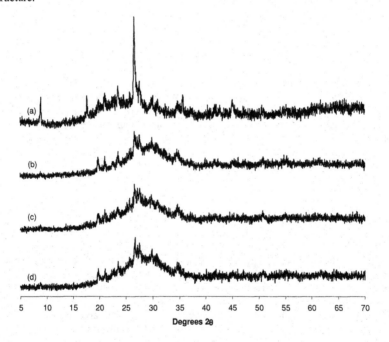

Figure 1. Cu-Kα X-ray diffractograms of (a) metakaolin, and (b-d) metakaolin/sodium silicate geopolymers with superficial Si:Al:Na molar ratios (b) 1.65:1.0:1.0, (c) 2.15:1.0:1.0, (d) 2.65:1.0:1.0, cured in sealed plastic bags at 40°C for 24h.

The broad hump at 20-25° 2θ shown in diffractogram (a) of Figure 1 is characteristic of metakaolin. The other peaks in this diffractogram are due to a muscovite impurity. Unfortunately, the XRD hump of metakaolin overlaps with that due to any high-silica gel regions present [7] and so this region cannot be considered a good measure of the degree of reaction, particularly in higher-silica systems such as that shown by diffractogram (d). Sometimes described as a 'diffuse halo peak,' the broad hump at 28° 2θ in diffractograms (b) to (d) of Figure 1 is generally attributed to the amorphous aluminosilicate gel assumed by most authors to be the primary binder phase present in geopolymeric systems. However, high-resolution microscopy of geopolymeric systems has recently shown the binder phase to be comprised largely of nanosized particles [8, 9]. Having previously noted that small crystallites produce broadened diffraction peaks, the possibility that the presence of the characteristic geopolymer hump is intrinsically connected to the presence of crystallinity in the observed particulate component of the geopolymeric binder phase must be considered seriously.

The investigation of Yang et al. [10] provides an initial basis for the identification of this peak. These investigators mixed sodium aluminate with colloidal silica and heated the mixture to give a mixture of faujasite and Na-A zeolites, obtaining XRD diffractograms at regular stages throughout the reaction. The initial broad peak at ~22° 2θ attributable to the presence of colloidal silica with minor aluminate inclusions was initially converted to a second broad peak centred at 28° 2θ similar to that observed in geopolymers, and which sharpened into the distinct peaks characteristic of faujasite and Na-A zeolites upon further heating. This second broad peak is identified as being due to the X-ray amorphous precursor gel from which nucleation of zeolites occurs. It has also been shown that this gel contains the nanometre-sized crystallites which will act as nucleation centres for the zeolites to be formed, as will be further discussed later in this paper. It is therefore entirely plausible that these crystallites contribute in a significant way to the diffraction pattern of the gel as a whole. The fact that the broad peak is observed to sharpen into the characteristic zeolite peaks upon crystallization further supports this proposition, as the process of crystal growth maintains the structure of the 'quasicrystalline' nuclei while removing any size-broadening effects.

It is therefore clear that the broad ~28° 2θ peak is attributable to the initial development of crystalline zeolites on a length scale below the detection limits of XRD. The authors of the original study described this phase as 'precrystallization', with zeolite structure present on a length scale of no more than 4 unit cells [10], or approximately 8-10 nm. The identification of the corresponding broad 28° 2θ peak in geopolymeric systems as being due to the nanocrystalline phases identifiable by electron diffraction but amorphous to XRD [8] is therefore further supported.

A similar peak-sharpening effect was noted by Zhan et al. [11] during synthesis of nanometre-sized crystals of zeolite X. A broad peak centred at ~29° 2θ and attributed to nuclei of zeolite X was seen to sharpen in the diffractogram of the final product as the crystals grow above the detection limit of XRD. These authors showed that the broad peak due to crystallization on length scales below 10nm resolves into slightly sharper but still size-broadened peaks upon further crystallization, with a final particle size of 23±4 nm [11] calculated. This places an upper bound on the size of the crystallites observable in geopolymers, which do not show diffraction peaks as sharp as those observed by Zhan et al. in their final 23±4 nm product and must therefore be smaller than this.

The work of Dutta et al. [12] is also valuable in confirming the assignment of the broad 'amorphous hump' centred at ~28-29° 2θ to nanometre-sized zeolite nuclei. These authors compared the XRD and Raman spectroscopic data shown in Figure 2, obtained following treatment of a mixture of colloidal silica and sodium aluminate under conditions described in Table 1. It was concluded that the change in XRD peak position and shape with heating from diffractograms (a) and (b) to diffractogram (c) in Figure 2 was due to formation of zeolitic nuclei. Diffractograms (a) and (b) in Figure 2 may be considered to represent vacuum-dried silica gel with slight (<8%) Al incorporation in the tetrahedral gel network. The peaks in these diffractograms occur at a 2θ angle only slightly lower than those attributable to zeolitic nuclei, but the significant difference in composition between these high-silica gels (Table 1) and geopolymerization systems allows identification of diffractogram (c) in Figure 2 as the diffractogram most likely to be comparable to the nanocrystalline portion of geopolymeric materials. However, the presence of relatively high-silica gel regions as a matrix in which the crystalline sections of geopolymers are embedded is definitely a proposition worthy of further investigation. Formation of such a phase will be incorporated into the reaction kinetic model to be developed later in this paper.

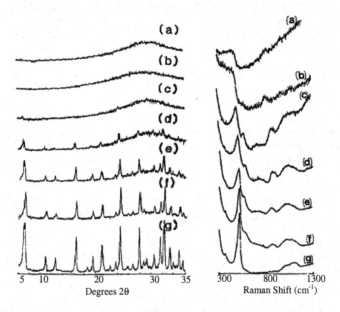

Figure 2. Cu-Kα X-ray diffractograms and Raman spectra of the solid phase present at different times during aging and crystallization of a colloidal silica/sodium aluminate mixture, as described in Table 1. Reprinted with permission from reference [12]. Copyright 1987 American Chemical Society.

Table 1. Preparation regimes for samples used in Figure 2 [12]

Sample	Preparation regime	Si/Al ratio in solids
(a)	Mixing of reactants (35 min)	26
(b)	Mixing + aging (room temperature, 24h)	15
(c)	Mixing + aging + heating at 90°C for 6h	2.5
(d)	Mixing + aging + heating at 90°C for 12h	1.7
(e)	Mixing + aging + heating at 90°C for 18h	1.7
(f)	Mixing + aging + heating at 90°C for 25h	1.8
(g)	Mixing + aging + heating at 90°C for 30h	1.8

The onset of nucleation is obvious in the Raman spectra of Figure 2, where peaks attributable to a zeolitic structure have developed after 6 hours of heating despite the largely featureless nature of the corresponding diffractogram. The XRD diffractograms of Figure 2 show sharpening of the zeolite diffraction peaks from the broad nanocrystallite peak as crystal growth occurs. This provides further confirmation of the assignment of the characteristic geopolymer 'amorphous hump' centred at approximately 28° 2θ to diffraction from zeolite nanocrystals less than 10nm in size.

Akolekar et al. [13] presented a sequence of powder X-ray diffractograms detailing the transformation of metakaolin to zeolite X (faujasite) in mixed KOH/NaOH solution at 51°C. As in the case of zeolite synthesis from colloidal silica and sodium aluminate as shown in Figure 2 [12] and also in other investigations of zeolite formation by leaching of metakaolin [14], a broad peak centred at approximately 28° 2θ is seen to replace the initial 22° metakaolin peak early in the transformation of metakaolin to the zeolite product. Peaks characteristic of zeolite A are observed in addition to the 'amorphous hump' in the intermediate stages of the reaction. However, these peaks decrease in intensity as zeolite X, the preferred product under the relatively low temperatures used, is formed. This is in accordance with the accepted applicability of Ostwald's law of successive reactions to zeolite synthesis systems. Davidovits [15] and Benharrats et al. [16] each obtained corresponding results in the reaction of kaolin or metakaolin with NaOH at 150°C and 80°C respectively, with zeolite A formed initially in each case and hydroxysodalite increasingly prominent as the reaction continued.

GEOPOLYMERIZATION AS A HYDROTHERMAL MINERAL SYNTHESIS

Hydrothermal synthesis using calcined clays, particularly metakaolin (calcined kaolinite), has long been used in the production of low-silica zeolites. The physicochemical conditions under which zeolites are obtained from metakaolin are very similar to those used in geopolymerization. Temperature and water content are generally higher in zeolite syntheses than in geopolymerization, but there is no clear distinction between the conditions at which each product is obtained. An indication of the products obtained under different conditions is given in Figure 3, where the observed overlap between the different categories may be attributed to variations in other synthesis parameters including Si/Al ratio, the nature of the alkali cations present and the presence or absence of alkaline earth cations. Activation with alkali silicate rather than hydroxide solutions tends to give geopolymeric rather than fully crystalline zeolitic products at high temperature and low water content.

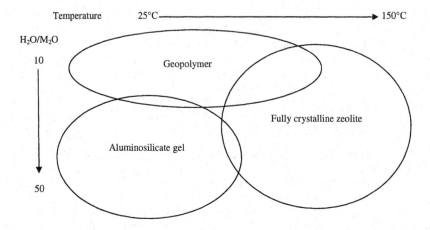

Figure 3. Temperature/composition regimes corresponding to different aluminosilicate products

Correlations between the mechanisms involved in geopolymerization and in zeolite synthesis have recently been strengthened with the publication by Yip *et al.* of TEM micrographs showing a semicrystalline to crystalline aluminosilicate geopolymer binder phase produced by activation of metakaolin with sodium silicate [8]. These authors showed that calcination of kaolinite to metakaolin leaves the fundamentally hexagonal kaolinite particle morphology largely unchanged, while dehydroxylating the clay and changing the coordination state of Al from Al(VI) to a mixture primarily containing Al(IV) and Al(V). Following geopolymerization, regions of two distinct aluminosilicate phases can be seen in addition to remaining undissolved hexagonal metakaolin particles. One of these phases can be seen by electron diffraction to be amorphous with possibly some small crystalline inclusions, and shows no regular particle morphology. The other is seen by electron diffraction to be polycrystalline, with a particle morphology that is approximately circular with a radius of 0.2μm [8]. However, no new crystalline phases were identifiable in XRD analysis of this geopolymeric product. It can be deduced from these seemingly conflicting results that the crystallinity observed in geopolymers by electron diffraction is present on a length scale below the detection limit of XRD instrumentation. Similar apparent discrepancies between XRD and electron diffraction results have been noted in studies of aluminosilicate zeolite precursor gels [17], as well as in a variety of other inorganic systems. In each case, crystallinity on a length scale of around 5nm was detectable by electron diffraction experiments but not by XRD.

From a thermodynamic viewpoint, the formation of crystalline structures has been described as the "victory of enthalpy over entropy" [6]. It is therefore expected that those geopolymeric systems that set most slowly will display a greater degree of crystallinity than the more rapidly-setting systems due to the increased opportunity for rearrangement into more energetically favourable structures. Also, geopolymeric materials utilized or stored in wet or humid environments, allowing some structural rearrangement to continue after setting, will display an increase in crystallinity with time. Such an increase has been observed by Palomo *et al.* [18], where metakaolin-based geopolymeric mortars that initially displayed complete X-ray amorphicity developed significant formation of faujasite zeolites when immersed in various aqueous environments including deionized water, seawater, 0.31M Na_2SO_4 and 1mM H_2SO_4 solutions. Recent work by Duxson *et al.* [19] has also shown significant development of crystalline faujasite when low-silica geopolymer samples are tightly sealed to prevent water loss and cured at 40°C for extended times.

The observed increase in crystallinity with time in such a wide variety of environments is in itself an important theoretical result relating to the fundamental nature of geopolymers. However, an even more critical insight may be obtained by combining this observation with the development of mechanical properties of geopolymers over time. The paper of Palomo *et al.* [18] also includes discussion of variation in flexural strength with time of exposure to the various aqueous environments studied. After an initial decrease attributable to loss of the more soluble components, strength is observed to increase with time of exposure to the aqueous environments for the 270 day period of the study. A more spectacular example of this phenomenon was noted by Glukhovsky [20] in the compressive strength development of slag-based geopolymers used to line irrigation channels in the former Soviet Union. These materials tripled in compressive strength over a 25 year period, without suffering any observable degree of corrosion. While this cannot be attributed directly to the development of crystallinity, the fact that strength and crystallinity appear to be at least interrelated, if not explicitly correlated, displays further the importance of developing an understanding of the crystallization chemistry of aluminosilicate systems if the performance and properties of geopolymeric materials are to be optimized.

ALUMINOSILICATE CRYSTALLIZATION MECHANISMS AND KINETICS

Hydrothermal crystallization in aluminosilicate systems of similar composition to geopolymer syntheses generally gives zeolitic products. Zeolites are a class of microporous, generally metastable aluminosilicate framework structures with widespread industrial application in catalysis, ion exchange and gas separation. The zeolitic phases commonly formed in geopolymeric systems include faujasites (zeolites X and Y), chabazite, gismondine (Na-P1) and zeolite A. Hydroxysodalite is also produced, and although technically a feldspathoid rather than a zeolite due to its relatively dense framework structure, this is largely a technical rather than a chemical distinction as its synthesis mechanisms are generally identical to those of 'true' zeolites. Hydroxysodalite is often considered in general discussions of zeolite chemistry, and will be classified along with the 'true' zeolites for the purposes of clarity in this study. Zeolite formation is favored at lower silicate modulus R_m (Na_2O/SiO_2 in activating solution) and higher temperature.

Zeolite synthesis procedures on both industrial and laboratory scales vary significantly depending on the exact nature of the product to be formed, particularly with regard to the presence or absence of a solid aluminosilicate source. Significant controversy regarding the exact mechanisms involved in zeolite crystallization has taken place over the past several decades, with the proliferation of different synthesis procedures only serving to add to the confusion. However, it has been conclusively proven by carefully-designed experimental techniques that at least two very distinct mechanisms will both give zeolitic products: solution-mediated transport and solid-solid transformation [21]. These mechanisms are represented schematically by Figure 4.

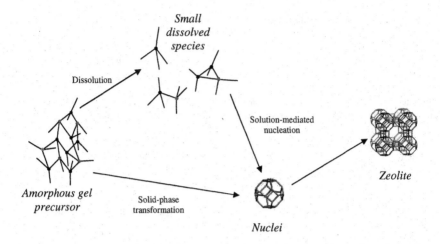

Figure 4. Schematic representation of solution-mediated and solid-solid transformation synthesis methods for crystalline aluminosilicates from an amorphous gel precursor.

Aiello *et al.* [22] carried out one of the first comprehensive microscopic studies of hydrothermal zeolite synthesis, and found that the initial particle morphology seen in the synthesis of zeolites in dilute solution is leaf-like, or 'lamellar'. The newly formed lamellae were initially amorphous to electron diffraction, but were seen by electron microscopy and diffraction to develop nanocrystalline regions within the general lamellar geometry as the reaction progressed. Subotić and co-workers [17] have since shown that the amorphous precursor gels from which zeolites are crystallized in fact contain many 'quasicrystalline' regions, which act as sites for zeolite nucleation as the gel dissolves. The effect of these regions in determining the synthesis kinetics of zeolites is expected to have significant correlation with the observed kinetics of geopolymerization, as it is these zeolitic nanocrystallites that are identified in this investigation as a highly significant structural component of geopolymers.

Nucleation in zeolite syntheses has been shown to occur via replacement of water by small aluminate or silicate species in the hydration shells of charge-balancing cations [21]. The rate of nucleation occurring in a given system depends on many factors, primarily the degree of supersaturation of the solution and the presence or otherwise of nucleation triggers. Nucleation triggers may include undissolved particles, newly-formed crystals or amorphous aluminosilicate particles, or any other solid-liquid interface. In a geopolymerization system, the degree of supersaturation and the number of nucleation triggers present are very high, so the nucleation rate would be expected to be very high. Autocatalysis has been noted in reaction kinetic investigations of zeolite crystallization and in aluminosilicate weathering with secondary mineralization. Similar effects must therefore be considered in any mechanistic and kinetic examination of geopolymerization, and are incorporated into the mathematical model to be developed later in this paper.

An early investigation of the formation of hydroxysodalite from quartz [23] found that very rapid nucleation, attributable to very high local silicate concentration in the aluminate solution near the quartz particle surfaces, led to formation of submicron-sized hydroxysodalite particles. Similar effects in geopolymerization would therefore be expected to be observed very close to the surfaces of metakaolin particles, where the release of high levels of $Al(OH)_4^-$ into the concentrated alkali silicate solution immediately surrounding the particles would likewise be expected to cause very rapid nucleation.

On the basis of this observation, the identification of nanocrystalline zeolitic materials as a significant component of geopolymeric materials is not unexpected. An explanation for the distinction between alkaline- and alkali silicate-activated geopolymeric systems is also now possible, with the differing rates of nucleation and crystal growth in each scenario playing a large part in determining the physicochemical and engineering properties of the products in each case. The observation of Rahier et al. [24] that crystallinity decreases as more silicate is added to the activating solution in geopolymerization can now be justified from a theoretical standpoint. This will be incorporated into the modeling of geopolymerization later in this paper by description of the competition for nutrients between the relatively high-silica amorphous gel phase and the relatively low-silica nanocrystalline zeolitic phases. Additionally, with nucleation not happening in such close proximity to the particle surfaces in hydroxide-activated as in silicate-activated geopolymers, the binding of the particles into the geopolymeric matrix is likely to be less strong. This provides a possible explanation for the observation that the mechanical strength of geopolymers formed by hydroxide activation is lower than in silicate activation [25], as will be further discussed later in this paper.

The investigation of Xu and van Deventer [26] into the geopolymerization of kaolinite/stilbite mixtures confirmed that the degree of binding of unreacted particles into the geopolymeric matrix plays a highly significant role in determining the strength of a geopolymer. In this investigation, it was shown that the majority of geopolymer samples fractured at the boundary between the binder phase and unreacted particles. The only exception was a sample with a very low degree of binder formation due to the low levels of reactive solid aluminosilicate and soluble silicate used, which was observed to fracture in the partially-formed binder phase. This may correspond to the observation of Phair et al. [27] that addition of any more than 3% zirconia by mass weakens a geopolymeric matrix by disrupting the formation of the binder phase, because the binder phase is primarily responsible for the strength development of a geopolymeric system.

The effect of the presence of Ca^{2+} on geopolymerization has recently been the subject of a number of detailed investigations [28, 29]. The conclusions of these investigations will not be repeated in detail here, other than to note that calcium silicate hydrate (CSH) compounds and $Ca(OH)_2$ precipitates have been observed in geopolymeric systems. The amount of Ca^{2+} added and the form in which it is added both play a significant role in determining the physical properties of the final geopolymer. The levels of dissolved silicate in the activating solution also plays a highly significant role in determining the effects of calcium by controlling the pH of the activating solution and therefore influencing the relative stabilities of the different calcium-containing precipitates.

It has been observed [27, 30] that CaO if present at levels above 3% by mass interferes with, but does not prevent, formation of XRD-observable crystalline zeolites. The formation of Ca-containing precipitates as observed by Yip and van Deventer [28] will provide a large number of potential nucleation sites at the solid-liquid boundaries thus formed. Nucleation at a proportion of these additional sites will then cause the total number of nuclei present to be higher than in the absence of calcium, leading to a smaller mean crystallite size and therefore lower observable aluminosilicate crystallinity. However, the formation of calcium silicate hydrates will also affect the formation of zeolite nuclei by the removal of a proportion of the excess silicate from solution [30], thereby reducing the supersaturation levels. This reduces the primary driving force for nucleation and crystal growth, and may therefore compete with the accelerating effects of added nucleation to further complicate description of the kinetic effects of calcium addition.

ANALYSIS OF EXISTING MECHANICAL STRENGTH RESULTS

Unfortunately, the most commonly employed measure of the success or otherwise of a geopolymerization process is the compressive strength of the final product. This is most likely due to the low cost and simplicity of compressive strength testing, and the importance of strength development as a primary measure of the utility of materials in different applications in the construction industry. Due to the wide variety of product sizes, geometries, strength testing apparatus and procedures used by different authors, strength results are generally not directly comparable across different investigations. Therefore, comparison of results or identification of trends by numerical comparison of data across different studies cannot be undertaken with any degree of accuracy.

Xu and van Deventer [31] tested a mixture of kaolinite, albite and fly ash which, when activated with alkali silicate solution, gave a stronger geopolymeric product than any combination of two of the three solid aluminosilicate sources used. This was attributed to the differing contributions of each of the source materials to the product: rapid solidification of highly soluble components leached from fly ash gave early strength, albite dissolved relatively little but due to its high hardness acted as an aggregate in the cementitious product, and kaolinite reacted slowly to give high final strength.

However, the results of investigations detailing the effects of large crystalline inclusions on the mechanical properties of geopolymeric systems must be interpreted very carefully when attempting to explain the effects of nanometre-scale crystallinity as proposed in the current work. These crystal nuclei are much too small to act effectively as aggregates, and so any data obtained from systems with significant aggregate effects will not necessarily be directly applicable.

Having outlined the limitations of mechanical strength data in the analysis of geopolymerization, it must also be observed that several highly significant results regarding chemical structure may be obtained from these data. In particular, the importance of the charge-balancing role of cations within a geopolymeric structure is exemplified by Figure 5 [32]. This plot shows a sharp maximum in both compressive and tensile strength at a Na/Al ratio of exactly 1, corresponding to a single Na^+ cation balancing the charge on each tetrahedral Al centre. Corresponding but less detailed results for sodium silicate-activated metakaolin have also been published by Rahier *et al.* [33].

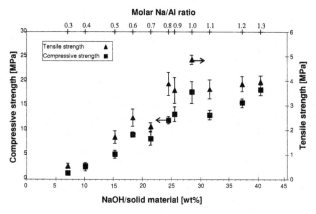

Figure 5. Influence of the Na/Al ratio on the mechanical properties of geopolymers formed by activation of metakaolin with NaOH solution. Data from reference [32].

The work of Rowles and O'Connor [25] appears initially to contradict this assertion, showing a peak in strength at Na/Al ~1.25. However, the formation of crystalline Na_2CO_3 during geopolymerization was also noted, and consumed a significant proportion of the Na used. EDS analysis of the geopolymeric binder phase of 'medium' and 'high' strength products with nominal Na/Al ratios of 1.29-1.5 in the same study showed actual Na/Al ratios of around 0.9 in the binder phase. It was suggested that this figure may be a slight under-reporting of the actual Na content due to instrumental difficulties [25], thereby confirming the charge-balancing requirement of Na/Al = 1. In contrast, the 'low' strength product in the same investigation, with nominal Na/Al 0.7, showed negligible formation of Na_2CO_3 and a binder-phase Na/Al ratio matching the nominal ratio exactly. This shows that the charge-balancing requirement of the negatively charged tetrahedral Al centres is stronger than the propensity of the system to form Na_2CO_3, and therefore that the charge-balancing positions will be filled in preference to carbonation in systems without sufficient alkali cations.

The requirement for a particular stoichiometric M^+/Al ratio to achieve maximum strength in geopolymeric binders adds further support to the proposal that these binders display a significant degree of chemical ordering, and nanocrystallinity in particular. All crystalline zeolitic structures previously mentioned as having been identified within geopolymeric binders require full charge-balancing by one alkali metal cation for each tetrahedral Al centre, as shown recently by Duxson et al. using MAS-NMR techniques [34]. In contrast, amorphous gel or glass structures do not show such strong charge-balancing requirements, as their less-ordered nature allows variation from strict tetrahedral geometry and therefore allows methods of charge compensation other than strict association of a single alkali metal cation with each Al centre.

ANALYSIS OF CALORIMETRIC DATA

A variety of data regarding the heat evolved during geopolymerization have been gathered by calorimetric techniques. The most important results obtained to date from calorimetric experiments generally fall into one of two categories: (1) determination of the correspondence between degree of reaction and physical properties, and (2) elucidation of the mechanism of reaction. The primary focus of this section will be the use of existing calorimetric data in conjunction with the nanocrystallinity hypothesis to describe the observed physical properties of geopolymers. Calorimetric data obtained from the literature will be used in the modeling section of this paper to allow validation of the proposed model, and to provide further insight into the reaction mechanisms postulated.

The data of Rahier et al. [33] presented in Figure 6 illustrate most clearly the relationship between degree of reaction and mechanical properties of a geopolymer. As all stages of the reaction between metakaolin and alkali silicate solution are exothermic, reaction enthalpy may be used as a direct representation of the extent of the reaction. Figure 6 shows a linear relationship, exact to within experimental error margins, between reaction enthalpy and product compressive strength. This corresponds with the absence of any observable aggregate effect in alkali-activated metakaolin systems due to the low hardness of metakaolin.

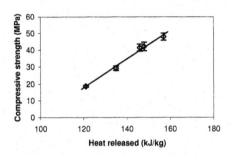

Figure 6. Relationship between total reaction enthalpy and product strength in geopolymers formed by activation of metakaolin with sodium silicate. Error bars represent one standard deviation. Data from reference [33].

In contrast, fly ash-based systems display no such clear general trend due to the compositional and physical differences between fly ashes from different sources or even between different batches of ash from the same source. This renders a detailed analysis of the general trends in these results extremely complex, and beyond the scope of this investigation.

Calorimetric and IR data have been used to show that the degree of crystallinity generally increases with alkali concentration in the NaOH activation of metakaolin/Ca(OH)$_2$ mixtures [29, 35]. This high alkalinity also tends to delay setting, particularly in the absence of dissolved silicates in the initial activating solution. It must be noted in the comparison of these data with those plotted in Figure 6 [33] that the strength/heat release relationship observed in Figure 6 is reversed entirely, with the data of Alonso and Palomo [35] showing that the samples with the highest heat release had the lowest strength. In addition, the setting of the samples of Alonso and Palomo was often very slow, suggesting that the very high alkalinity and therefore slow development of supersaturation of dissolved aluminate and silicate species led to a long 'induction period.' Formation of the more crystalline and therefore lower-energy species present in the systems that set slowest will therefore give greater total heat evolution than the rapid formation of less-crystalline species, but the slow setting gives no inherent strength advantages. The reversal of the strength/heat release relationship can therefore be largely attributed to the absence of dissolved silicate in the activating solution of Alonso and Palomo.

This contrast between the data of Rahier *et al.* [33] and those of Alonso and Palomo [35] provides additional justification for the explanations presented for observed differences in setting behavior and heat release between alkali silicate- and hydroxide-activated geopolymeric systems. The much lower silicate concentrations near dissolving particle surfaces in the hydroxide-activated systems of Alonso and Palomo will not lead to rapid nucleation in these regions. Therefore, those nuclei that do develop will face less competition for the nutrients required for their growth into crystals, providing the potential for them to develop into larger crystals than are possible in silicate-activated systems. This means that not only is the concept of nanocrystallinity in geopolymers highly plausible from a scientific standpoint, but it also has great potential importance in the prediction and explanation of the engineering properties of these materials.

PROPOSED REACTION KINETIC MODEL OF GEOPOLYMERIZATION

A schematic diagram of the chemical reaction sequence postulated to be responsible for the formation of geopolymers by activation of metakaolin with alkaline silicate solutions is presented in Figure 7. The proposed model in the current work is an extension of that presented by Faimon [36] in description of the weathering of aluminosilicate minerals in aqueous conditions. The model of Faimon allowed for dissolution of a primary mineral into aluminate and silicate monomers, association of these monomers via autocatalytic polymerization, and formation of an unidentified 'secondary mineral' phase. Its extension to geopolymerization is therefore relatively straightforward, requiring only incorporation of the effect of silicate oligomerization (species D) in the concentrated activator solutions, identification of the secondary mineral product G as the amorphous aluminosilicate component of the geopolymeric binder, and inclusion of a second pathway by which the zeolitic phases (Z) commonly observed in geopolymers are formed.

The abbreviations used in description of the reaction scheme are outlined in Table 2. All species are represented as molar quantities. The values k_1-k_8 are empirically determined rate constants. K_s represents the equilibrium between silicate monomers and the (inactive with respect to aluminate) larger silicate species formed in alkaline solution. This equilibrium is assumed to be maintained throughout the reaction, as the exchange reactions between silicate monomers and larger silicate species have been shown to occur on a timescale of less than 0.5s [37].

The reactions involved in the formation of geopolymers from metakaolin and sodium silicate solution, and the kinetic expressions used in the description of these reactions, are outlined in Table 3.

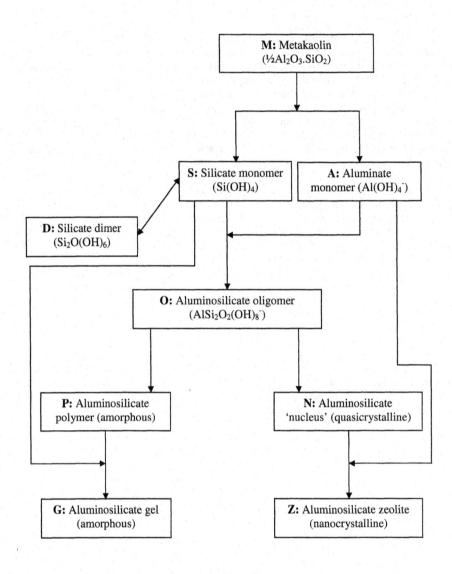

Figure 7. Schematic diagram of the reaction sequence of geopolymerization.

Table 2. Abbreviations used in description of the geopolymerization of metakaolin.

Symbol	Explanation
M	Metakaolin – assumed structural unit is ½(Al$_2$O$_3$.2SiO$_2$), to provide 1 unit each of Si and Al monomer upon dissolution
S	Silicate monomer – assumed structural unit is Si(OH)$_4$. Deprotonation neglected
A	Aluminate monomer, Al(OH)$_4^-$
O	Aluminosilicate oligomer AlSi$_2$O$_2$(OH)$_8^-$ (Ref. [36]). Deprotonation neglected
D	Silicate oligomer – unreactive. Represented by dimer Si$_2$O(OH)$_6$, the most prevalent larger silicate species under geopolymerization conditions. Deprotonation neglected
P	Amorphous aluminosilicate polymer formed primarily by autocatalytic combination of species O
N	'Zeolitic nuclei' formed from species O
G	Aluminosilicate (high-silica) gel formed by addition of silicate monomer to species P
Z	'Zeolitic phase' (relatively low-silica) formed by addition of silicate and aluminate monomers to species N
W	Water – consumed and produced by each reaction. Changes in water content are potentially highly significant due to the water-starved nature of the system

Table 3. Reactions modeled, with corresponding kinetic expressions

	Reaction	ΔH_r (kJ/mol)	Rate expression	Comment
(1)	M + 4W → S + A	-1099.8	$r_1 = k_1 M$	$k_1 \propto$ initial surface area – Ref. [5]
(2)	A + 2S → O + 2W	-285.3	$r_2 = k_2 AS^2$	Stoichiometry as in Ref. [36]
(3)	2O → P + W	-74.1	$r_3 = k_3 O^2$	Slow – provides P for (4) to start
(4)	2P + 2O → 3P + W	-74.1	$r_4 = k_4 P^2 O^2$	Autocatalytic – Ref. [36]
(5)	2O → N + W	-74.1	$r_5 = k_5 O^2$	
(6)	N + S + A → Z + 2W	-50	$r_6 = k_6 NSA$	
(7)	2S ↔ D + W	-41.8		Assumed instantaneous
(8)	P + S → G + W	-50	$r_8 = k_8 PS$	

The enthalpy of dissolution of metakaolin in alkaline solution was calculated from standard heats of formation in addition to previously published data on the enthalpy of dissolution of metakaolin in HF [38]. The enthalpy of formation of species O and P were estimated based on data presented in reference [39] for similar species. In the absence of existing data, it has been assumed that the enthalpy of formation of species N is the same as that of species P. The heats of formation of species Z and G from their respective precursors have been estimated to be approximately -50 kJ/mol. The equilibrium relationship used to describe the formation of the silicate dimer was based on available literature data [40].

The influence of the initial metakaolin particle size is incorporated into the model by the use of a correction factor in the metakaolin dissolution rate expression. The rate constant k_1 is directly proportional to the initial surface area. Similar behavior was also observed by Rahier *et al.* [5], and this relationship will be seen in the following section of this paper to provide a generally accurate description of the effect of metakaolin particle size on the kinetics of geopolymerization.

The participation of water in the reaction scheme is modeled by a modifying factor in the rate constant of each reaction in which water is involved. The water content of the system is calculated at every timestep. Generation or consumption of water affects the rate of each reaction in ways that may initially appear contradictory to the stoichiometry of the reactions. These relationships and the reasoning behind them may be briefly summarized as follows:

- The rate of dissolution of metakaolin decreases in the presence of additional water. This is because the rate of aluminosilicate mineral dissolution in highly alkaline solutions is primarily controlled by a_{OH^-}, the activity of hydroxide ions. This activity is approximately proportional to [OH$^-$], so adding water to the system will decrease the hydroxide activity and therefore the dissolution rate.

- The rate of each condensation reaction increases in the presence of additional water under the severely water-depleted conditions observed in geopolymer synthesis. This seemingly counterintuitive statement may be justified by comparing the water content of the system to the water required for hydration of each cation. In a standard geopolymer synthesis, the ratio H_2O/Na_2O is approximately 10, meaning that approximately 5 water molecules are present for every alkali metal cation. The first-shell hydration number of the Na$^+$ cation is believed to be approximately six, so to fill the first hydration shells of the cations present will require more water than is present in the system. To account for this water-deficiency, the cations will instead bind strongly to dissolved anionic species, particularly aluminate and silicate monomers and/or small oligomers – species A, S and O in Table 2. This then stabilizes the anions, reducing the rate of reaction. When additional water is provided, the binding between the alkali metal cations and the small anions is slightly weakened, thereby accelerating the condensation reactions.

MODEL APPLICATION

Figure 8 shows a comparison between the heat flow predictions of the model presented in the previous section of this paper and the experimental modulated DSC data of Rahier et al. [5]. From the graphs in Figure 8, it may be observed that the model predictions fit the experimental data exceptionally well for surface areas of 16 and 13 m^2/g (Figures 8b and c) and satisfactorily for the metakaolin of specific surface area 20 m^2/g (Figure 8a). The fit for the largest particle size, with a specific surface area of 9 m^2/g, is less impressive. However, the fact that the model generally overestimates the reaction rate at the largest particle size tested may suggest that the model is not completely accurate in the way particle size is accounted for in calculations. This is to be expected, as a simple correction in the dissolution rate constant will not accurately account for all the physical and chemical effects of the presence of larger particles. There may also be some degree of unreactive material associated with this size fraction due to the sedimentation process used to separate the particle size fractions. Hindered mass transport in the reaction of larger particles was noted by Rahier et al. [5] in this reaction system. A full description of the kinetics of geopolymerization must incorporate such effects on a fundamental level rather than the empirical approach utilized here, but such work is beyond the scope of the current investigation.

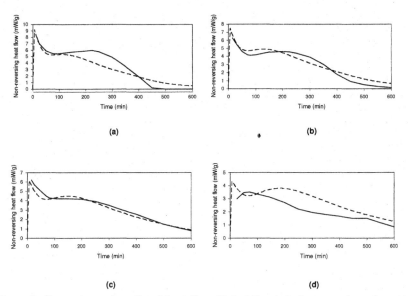

Figure 8. Comparison of predicted heat flow (dotted lines) during geopolymeric setting to experimental modulated DSC data of Rahier *et al.* [5] (solid lines) for initial metakaolin surface areas of (a) 20, (b) 16, (c) 13 and (d) 9 m^2/g. Reaction temperature in all cases was 35°C, with reaction mix composition $Na_2O/Al_2O_3 = 1.0$, solution $SiO_2/Na_2O = 1.4$, $H_2O/Na_2O = 10$.

CONCLUSIONS

From existing published experimental data and by comparison with related hydrothermal mineral synthesis systems, it may be concluded that a significant component of the binder phase formed in geopolymerization is nanometre-sized crystalline structures resembling the nuclei around which zeolites crystallize. Agglomeration of these nanocrystallites with other less-ordered aluminosilicate structures forms a high-performance mineral binder, commonly referred to as a 'geopolymer.' Unreacted particles of whatever solid aluminosilicate source was used in geopolymerization will be bound within this matrix. Unreactive aggregate particles such as zirconia are not chemically bound into the aluminosilicate phase, and therefore may give decreased strength if used in excessive quantities.

The use of alkali silicate activator solutions gives a product with higher strength than if alkali hydroxides are used, and this is attributable to the rapid nucleation of solid products immediately surrounding the dissolving aluminosilicate source particles. These form a matrix binding the geopolymeric product together, and reduce the likelihood of the meso- to macroscopic flaws responsible for the low strength of some geopolymers.

Powder X-ray diffraction analysis provides valuable insight into the structures present in geopolymers, but is complicated greatly by the nanosized nature of the bulk of the crystalline products present. Electron diffraction is able to identify these phases as crystalline, but to date has not been applied as a quantitative technique in the analysis of geopolymers. Mechanical strength testing is not an ideal means of determining the success or otherwise of geopolymerization on a nanostructural level, but if carried out correctly is capable of providing data of some value in structural analyses. Calorimetric data is potentially of great importance in the evaluation of geopolymerization, but must be analysed carefully if all possible effects are to be taken into account. The identification of zeolitic nanocrystals within the geopolymeric binder is potentially a highly significant observation, as it provides a link between the chemical composition and engineering properties of geopolymeric materials, thereby providing the potential for tailoring of the properties of geopolymers for desired applications based on an understanding of their nanometre-level structure.

A reaction kinetic model is presented to describe the formation of aluminosilicate geopolymers by sodium silicate activation of metakaolin. Incorporation of thermodynamic data from the literature allows comparison with experimental calorimetric results, showing that the model accurately predicts the heat flow throughout the reaction process under a range of activator compositions and with several different initial metakaolin particle sizes.

ACKNOWLEDGEMENTS

The authors wish to thank Peter Duxson for generously supplying the XRD data used in Figure 1. This work was funded through the Particulate Fluids Processing Centre (PFPC), a Special Research Centre of the Australian Research Council, and through an Australian Postgraduate Award scholarship awarded to JLP.

REFERENCES

[1] J. Davidovits, "Geopolymers - Inorganic polymeric new materials," *Journal of Thermal Analysis*, **37**(8) 1633-1656 (1991).

[2] H. Xu and J.S.J. van Deventer, "The geopolymerisation of alumino-silicate minerals," *International Journal of Mineral Processing*, **59**(3) 247-266 (2000).

[3] P.V. Krivenko, "Alkaline cements," in *Proceedings of the First International Conference on Alkaline Cements and Concretes*. Kiev, Ukraine: VIPOL Stock Company: 11-129. P.V. Krivenko, Ed. (1994)

[4] S. Swier, G. van Assche, A. van Hemelrijck, H. Rahier, E. Verdonck and B. van Mele, "Characterization of reacting polymer systems by temperature-modulated differential scanning calorimetry," *Journal of Thermal Analysis*, **54**(2) 585-604 (1998).

[5] H. Rahier, J.F. Denayer and B. van Mele, "Low-temperature synthesized aluminosilicate glasses. Part IV. Modulated DSC study on the effect of particle size of metakaolinite on the production of inorganic polymer glasses," *Journal of Materials Science*, **38**(14) 3131-3136 (2003).

[6] G.R. Desiraju, "In search of clarity," *Nature*, **423** 485 (2003).

[7] L.A. García-Cerda, O. Mendoza-González, J.F. Pérez-Robles and J. González-Hernández, "Structural characterization and properties of colloidal silica coatings on copper substrates," *Materials Letters*, **56**(4) 450-453 (2002).

[8] C.K. Yip, G.C. Lukey and J.S.J. van Deventer, "Effect of blast furnace slag addition on microstructure and properties of metakaolinite geopolymeric materials," *Ceramic Transactions*, **153** 187-209 (2003).

[9] T.W. Cheng and J.P. Chiu, "Fire-resistant geopolymer produced by granulated blast furnace slag," *Minerals Engineering*, **16**(3) 205-210 (2003).

[10] S. Yang, A. Navrotsky and B.L. Phillips, "In situ calorimetric, structural, and compositional study of zeolite synthesis in the system $5.15Na_2O-1.00Al_2O_3-3.28SiO_2-165H_2O$," *Journal of Physical Chemistry B*, **104**(25) 6071-6080 (2000).

[11] B.-Z. Zhan, M.A. White, M. Lumsden, J. Mueller-Neuhaus, K.N. Robertson, T.S. Cameron and M. Gharghouri, "Control of particle size and surface properties of crystals of NaX zeolite," *Chemistry of Materials*, **14**(9) 3636-3642 (2002).

[12] P.K. Dutta, D.C. Shieh and M. Puri, "Raman spectroscopic study of the synthesis of zeolite Y," *Journal of Physical Chemistry*, **91**(9) 2332-2336 (1987).

[13] D. Akolekar, A. Chaffee and R.F. Howe, "The transformation of kaolin to low-silica X zeolite," *Zeolites*, **19**(5-6) 359-365 (1997).

[14] J. Rocha, J. Klinowski and J.M. Adams, "Synthesis of zeolite Na-A from metakaolinite revisited," *Journal of the Chemical Society - Faraday Transactions*, **87**(18) 3091-3097 (1991).

[15] J. Davidovits, "Structural characterization of geopolymeric materials with X-ray diffractometry and MAS-NMR spectrometry," in *Geopolymer '88 - First European Conference on Soft Mineralurgy*. Compeigne, France: Universite de Technologie de Compeigne: 149-166. J. Davidovits and J. Orlinski, Eds. (1988)

[16] N. Benharrats, M. Belbachir, A.P. Legrand and J.-B. D'Espinose de la Caillerie, "^{29}Si and ^{27}Al MAS NMR study of the zeolitization of kaolin by alkali leaching," *Clay Minerals*, **38**(1) 49-61 (2003).

[17] B. Subotiæ, A.M. Tonejc, D. Bagoviæ, A. Èižmek and T. Antoniæ, "Electron diffraction and infrared spectroscopy of amorphous aluminosilicate gels," in *Zeolites and Related Microporous Materials: State of the Art 1994*: Elsevier Science: 259-266. W. Holderich, Ed. (1994)

[18] A. Palomo, M.T. Blanco-Varela, M.L. Granizo, F. Puertas, T. Vazquez and M.W. Grutzeck, "Chemical stability of cementitious materials based on metakaolin," *Cement and Concrete Research*, **29**(7) 997-1004 (1999).

[19] P. Duxson, G.C. Lukey and J.S.J. van Deventer, "Thermal stability of metakaolin-based geopolymers," *Manuscript in preparation* (2004).

[20] V.D. Glukhovsky, "Ancient, modern and future concretes," in *Proceedings of the First International Conference on Alkaline Cements and Concretes*. Kiev, Ukraine: VIPOL Stock Company: 1-9. P.V. Krivenko, Ed. (1994)

[21] D.P. Serrano and R. van Grieken, "Heterogeneous events in the crystallization of zeolites," *Journal of Materials Chemistry*, **11**(10) 2391-2407 (2001).

[22] R. Aiello, R.M. Barrer and I.S. Kerr, "Stages of zeolite growth from alkaline media," in *Molecular Sieve Zeolites*, L.B. Sand, Ed. American Chemical Society: Washington. 44-50. (1971)

[23] R.A. Cournoyer, W.L. Kranich and L.B. Sand, "Zeolite crystallisation kinetics related to dissolution rates of quartz reactant," *Journal of Physical Chemistry*, **79**(15) 1578-1581 (1975).

[24] H. Rahier, W. Simons, B. van Mele and M. Biesemans, "Low-temperature synthesized aluminosilicate glasses. 3. Influence of the composition of the silicate solution on production, structure and properties," *Journal of Materials Science*, **32**(9) 2237-2247 (1997).

[25] M. Rowles and B. O'Connor, "Chemical optimisation of the compressive strength of aluminosilicate geopolymers synthesised by sodium silicate activation of metakaolinite," *Journal of Materials Chemistry*, **13**(5) 1161-1165 (2003).

[26] H. Xu and J.S.J. van Deventer, "Microstructural characterisation of geopolymers synthesised from kaolinite/stilbite mixtures using XRD, MAS-NMR, SEM/EDX, TEM/EDX, and HREM," *Cement and Concrete Research*, **32**(11) 1705-1716 (2002).

[27] J.W. Phair, J.S.J. van Deventer and J.D. Smith, "Mechanism of polysialation in the incorporation of zirconia into fly ash-based geopolymers," *Industrial & Engineering Chemistry Research*, **39**(8) 2925-2934 (2000).

[28] C.K. Yip and J.S.J. van Deventer, "Microanalysis of calcium silicate hydrate gel formed within a geopolymeric binder," *Journal of Materials Science*, **38**(18) 3851-3860 (2003).

[29] M.L. Granizo, S. Alonso, M.T. Blanco-Varela and A. Palomo, "Alkaline activation of metakaolin: Effect of calcium hydroxide in the products of reaction," *Journal of the American Ceramic Society*, **85**(1) 225-231 (2002).

[30] P. Catalfamo, S. Di Pasquale, F. Corigliano and L. Mavilia, "Influence of the calcium content on the coal fly ash features in some innovative applications," *Resources, Conservation and Recycling*, **20**(2) 119-125 (1997).

[31] H. Xu and J.S.J. van Deventer, "Geopolymerisation of multiple minerals," *Minerals Engineering*, **15**(12) 1131-1139 (2002).

[32] C. Kaps and A. Buchwald, "Property controlling influences on the generation of geopolymeric binders based on clay," in *Geopolymers 2002. Turn Potential into Profit.* Melbourne: University of Melbourne: CD-ROM Proceedings. G.C. Lukey, Ed. (2002)

[33] H. Rahier, B. van Mele, M. Biesemans, J. Wastiels and X. Wu, "Low-temperature synthesized aluminosilicate glasses. 1. Low-temperature reaction stoichiometry and structure of a model compound," *Journal of Materials Science*, **31**(1) 71-79 (1996).

[34] P. Duxson, G.C. Lukey and J.S.J. van Deventer, "Multinuclear MAS-NMR investigation of geopolymerisation," *Solid State Nuclear Magnetic Resonance*, **Submitted** (2004).

[35] S. Alonso and A. Palomo, "Alkaline activation of metakaolin and calcium hydroxide mixtures: influence of temperature, activator concentration and solids ratio," *Materials Letters*, **47**(1-2) 55-62 (2001).

[36] J. Faimon, "Oscillatory silicon and aluminum aqueous concentrations during experimental aluminosilicate weathering," *Geochimica et Cosmochimica Acta*, **60**(15) 2901-2907 (1996).

[37] E.K.F. Bahlmann, R.K. Harris, K. Metcalfe, J.W. Rockliffe and E.G. Smith, "Silicon-29 NMR self-diffusion and chemical-exchange studies of concentrated sodium silicate solutions," *Journal of the Chemical Society - Faraday Transactions*, **93**(1) 93-98 (1997).

[38] M. Murat and M. Driouche, "Conductimetric investigations on the dissolution of metakaolinite in dilute hydrofluoric acid. Structural implications.," *Clay Minerals*, **23**(1) 55-67 (1988).

[39] C.R.A. Catlow, A.R. George and C.M. Freeman, "Ab initio and molecular-mechanics studies of aluminosilicate fragments, and the origin of Lowenstein's rule," *Chemical Communications*(11) 1311-1312 (1996).

[40] P. Caullet and J.L. Guth, "Observed and calculated silicate and aluminosilicate oligomer concentrations in alkaline aqueous solutions," in *Zeolite Synthesis*, M.L. Occelli and H.E. Robson, Eds. American Chemical Society: Washington DC. 83-97. (1989)

MICROSTRUCTURAL CHARACTERISATION OF METAKAOLIN-BASED GEOPOLYMERS

Peter Duxson
Department of Chemical and Biomolecular
Engineering
The University of Melbourne
Victoria 3010, AUSTRALIA.

Seth W. Mallicoat
Department of Materials Science
and Engineering
University of Illinois at Urbana-Champaign
1304 W. Green St.,
Urbana IL 61801, USA

Grant C. Lukey
Department of Chemical and Biomolecular
Engineering
The University of Melbourne
Victoria 3010, AUSTRALIA.

Waltraud M. Kriven
Department of Materials Science
and Engineering
University of Illinois at Urbana-Champaign
1304 W. Green St.,
Urbana IL 61801, USA

Jannie S.J. van Deventer
Department of Chemical and Biomolecular Engineering
The University of Melbourne
Victoria 3010, AUSTRALIA.

ABSTRACT

Inorganic polymers (geopolymers) are formed by the alkali silicate activation of aluminosilicate materials such as metakaolin. Given correct mix-design and processing, geopolymer matrices can exhibit mechanical properties comparable or better than current cement systems, which makes them suited to a variety of construction applications. However, the variation in published mechanical and thermal properties is often a source of concern about the commercial and industrial maturity of geopolymeric materials. Experimental variations are often a result of inappropriate sample preparation, and poor quantification of system parameters. The present study details an appropriate nomenclature and standard for sample preparation to be used in geopolymeric research. These concepts are illustrated with the use of SEM to clarify the different microstructures that result from changes in the concentration of silicate and alkali type (i.e. KOH or NaOH) in the activating solution of metakaolin-based geopolymers. It is found that by increasing silicate concentration in the activating solution, the homogeneity of the bulk microstructure is improved. In all samples there is an observed increase in porosity close to unreacted metakaolin particles. It is proposed that this phenomenon is due to the nucleated growth of the geopolymer gel from the bulk to the particle surface. The macro-scale properties of these materials are correlated to each microstructure, and the results indicate that there is an optimal silicate concentration and type of alkali to use in formulation development.

INTRODUCTION

Davidovits coined the term 'geopolymer to' describe aluminosilicate binders formed through alkali silicate activation of aluminosilicate materials[1]. Geopolymers are often confused with alkali-activated cements, which were originally developed by Glukhovsky in the Ukraine during the 1950's[2]. Glukhovsky worked predominantly with alkali activated slags containing large amounts of calcium, compared to Davidovits, who pioneered calcium free systems formed from calcined clays. The purely alkali aluminosilicate framework of a geopolymeric binder is intrinsically fire resistant and has been shown to have excellent thermal stability far in excess of traditional cements[3]. Geopolymers have also been shown to exhibit comparable mechanical properties to those of Ordinary Portland Cement (OPC)[4, 5].

The present understanding of the molecular structure of geopolymers is currently undergoing a fundamental reappraisal with a greater emphasis being placed on determining the framework structure of the aluminosilicate matrix. As such, the polysialate nomenclature developed by Davidovits[1] for describing the molecular structure of geopolymers is also under review. The term sialate is an abbreviation for silicon-oxo-aluminate and is used to describe the bonding of silicon and aluminium by a bridging oxygen. Polysialates are said to have an empirical formula of:

$$M_n \{(SiO_2)_z AlO_2\}_n, wH_2O$$

where M is an alkali cation (eg. Na^+ or K^+), n is the degree of polycondensation and z is either 1,2 or 3. The three polysialate oligomers, poly(sialate) (PS), poly(sialate-siloxo) (PSS) and poly(sialate-disiloxo) (PSDS), are described as chain and ring polymers with Si^{4+} and Al^{3+} in IV-fold coordination with oxygen and range from amorphous to semi-crystalline. The one-dimensional polysialate oligomers fail to describe the three-dimensional environment of each silicon and aluminium centre in the binder and thus have little value in describing the molecular structure of geopolymeric binders. No evidence or method has been presented in the literature to support the existence of the polysialate building blocks. Furthermore, none of the geopolymer literature makes use of these proposed polysialate oligomers in reference to the molecular structure of geopolymers. The polysialate nomenclature has only been adopted in a practical sense for describing the superficial Si:Al ratio of geopolymers in research publications, where PS, PSS and PSDS describe systems with Si:Al ratios of 1,2 and 3 respectively. The standard Q-species notation for describing silica and alumina centres in tetrahedral coordination is becoming more prevalent in geopolymer science. The Q-species notation first described by Engelhardt[6] is used throughout the aluminosilicate literature and fully describes the entire aluminosilicate framework in terms of proportions of each $Q^4(mAl)$ species (where $0 \leq m \leq 4$).

The majority of published studies on geopolymer systems have focussed on determination of bulk mechanical properties, namely the ultimate compressive strength. While variations in raw materials, such as the effect of different calcium containing raw-materials[7, 8], additives[9], curing conditions[10] and post-curing chemical treatments[11] on strength have been investigated in some depth, few other relevant mechanical properties have been measured in these studies, including density and Young's modulus, which are of particular relevance in architectural applications.

One of the major drawbacks in geopolymer research is the lack of a standardised experimental composition and processing technique. Each study utilises different solid raw materials, activating solutions, curing conditions, sample geometries and naming conventions making comparison of experimental data difficult and qualitative at best. Recent studies have begun to address this problem, basing all experiments on a systematic series of metakaolin-based geopolymers[12]. Within this system a wide variety of Si:Al ratios and alkali composition are investigated by variation of the activating solution composition. Differential Scanning Calorimetry (DSC) has been utilised by Rahier et al. to show that the heat of reaction during geopolymerisation is inherently linked to the ratio of Al:M (where M is an alkali cation). The heat of reaction increases until the Al:M ratio reaches unity, beyond which no further heat is released. Furthermore, a maximum in both compressive and tensile strength of metakaolin-based geopolymers is found when Al:M = 1. It can be inferred from these results that only enough alkali is required in the reaction mixture to balance aluminium in the raw materials. Despite this, recent experiments found the optimum strength of metakaolin-based geopolymers by simultaneous variation of both Al:M and Si:Al ratios[14]. It was noted that strength and microstructure varied with Si:Al and Al:M ratios, but they were unable to separate clearly the effect of both ratios since neither was held constant. The effect of Si:Al ratio on strength and microstructure should therefore be independently investigated.

The most common method for varying the composition of geopolymers based on the same solid aluminosilicate source is by manipulation of the activating solution. The concentration of dissolved silicon in the activating solution is measured as SiO_2/M_2O, where M is the sum of alkali cations. Therefore, when the Al_2O_3/M_2O ratio of a geopolymeric mix is used to determine the amount of alkali silicate activator required, the SiO_2/M_2O ratio of the activating solution can be used directly to measure an accurate Si:Al ratio. Si:Al ratio is used in preference to SiO_2/Al_2O_3 in order to avoid confusion since aluminium is not present in octahedral coordination. Aluminium within a geopolymeric binder is in tetrahedral geometry, which requires a charge balancing cation to maintain electronic neutrality. Manipulation of the silicon content of geopolymers by this method is, however, restricted by the physical limits of silicon solubility in alkali solutions. Typical H_2O/M_2O ratios of approximately 11 are limited to $SiO_2/M_2O \leq 2$. Therefore the superficial Si:Al ratio of a metakaolin-based geopolymer with Al:M = 1 is limited to $1 \leq Si:Al \leq 2$. Fumed silica has been added to the solid raw material component of geopolymers to artificially increase the Si:Al ratio beyond that obtainable in the activating solution, however samples exhibit evidence of unreacted silicon in the ^{29}Si MAS-NMR spectra and are categorised as poorly reacted by the authors.[15,16] The alkali composition of the activating solution can be easily varied by use of alkali hydroxides (ie. NaOH and KOH).

Initial studies of geopolymer microstructure have focused on the existence and distribution of unreacted particles and the chemical composition of newly formed binder in systems synthesised from multi-component materials, such as blast furnace slag and fly-ash.[8,17,18] A recent TEM study identified the nanoparticulate nature of geopolymeric binder.[19] Furthermore, geopolymers have been shown to have a nanoporous framework, with the characteristic pore size being dependant on the alkali cation[20]. Studies of fly-ash based systems identified quartz and mullite particles that act as aggregates in the final matrix, with no evidence of unreacted glassy aluminosilicates. It is therefore thought that the glassy material acts as the only source of aluminium and silicon for the binder in these systems. In contrast, fracture surface analysis of clay-based systems exhibit sheets of unreacted particles trapped in the binder[18]. Aluminosilicate particles capable of dissolving during reaction being present in the cured geopolymer indicate that the samples harden prior to complete reaction[18,21]. The particle size of metakaolin has been shown to affect significantly the rate and extent of dissolution during geopolymerisation[22]. Synthetic raw materials with accurately known Si:Al ratio, surface area and particle size could provide the basis for well-controlled synthesis of geopolymers. Recently investigations into geopolymers based on a synthetic raw material formed from melt quenched glass have been carried out, however phase separation during quenching created inert particles that were detected in the hardened samples by SEM.[23]

There is a clear need to understand the effect of different Si:Al ratios and alkali composition on the mechanical, physical and microstructural properties of geopolymers. This study investigates the link between the microstructure of metakaolin-based geopolymers and their respective mechanical and physical properties of ultimate compressive strength, Young's modulus and density. The compositions of geopolymers are formulated to ensure that the Al:M ratio is constant at unity, providing sufficient alkali to complete charge balancing of aluminium. The composition of the geopolymers studied is controlled by varying the composition of the activating solutions as described earlier. The microstructures of geopolymers are characterised both qualitatively by SEM and with quantitative Energy Dispersive X-ray (EDX) chemical analysis. The microstructure of geopolymers is then correlated with basic macro-scale physical properties of ultimate compressive strength, Young's modulus and superficial density. A mechanistic model for describing the formation of the microstructure of geopolymers is proposed.

EXPERIMENTAL METHODS

Materials

Metakaolin was purchased under the brand name of Metastar 402 from Imerys Minerals, UK. The metakaolin contains a small amount of muscovite as impurity. The chemical composition of metakaolin determined by X-Ray Fluorescence (XRF) was $2.3 \cdot SiO_2.Al_2O_3$. The Brunauer-Emmett-Teller (BET) surface area of the metakaolin, as determined by nitrogen adsorption on a Micromeritics ASAP2000 instrument, is $12.7 \text{ m}^2/\text{g}$, and the mean particle size (d_{50}) is $1.58 \mu m$.

Alkaline silicate solutions based on three differing ratios of alkali metal Na/(Na+K) = M (0.0, 0.5 and 1.0) with composition SiO_2/M_2O = R (0.0, 0.5, 1.0, 1.5 and 2.0) and $H_2O/M_2O = 11$ were prepared by dissolving amorphous silica in appropriate alkaline solutions until clear. Solutions were stored for a minimum of 24 hours prior to use to allow equilibration.

Geopolymer Synthesis

Geopolymer samples were prepared by mechanically mixing stoichiometric amounts of metakaolin and alkaline silicate solution with $Al_2O_3/M_2O = 1$ to form a homogenous slurry. After 15 minutes of mechanical mixing the slurry was vibrated for a further 15 minutes to remove entrained air before being transferred to Teflon moulds and sealed from the atmosphere. Samples were cured in a laboratory oven at 40°C and ambient pressure for 20 hours before being transferred from moulds into sealed storage vessels at ambient temperature and pressure for 2 weeks prior to use in NMR experiments. Samples synthesised from alkaline activator solutions containing only sodium cations will be referred to as Na-geopolymers, with samples made from sodium and potassium (1:1) and samples containing only potassium cations will be referred to as NaK- and K-geopolymers respectively. Five Na-, NaK- and K-geopolymers were each synthesised with different Si:Al ratios by use of the five different concentrations of alkali activator solutions, R = 0.0, 0.5, 1.0, 1.5 and 2.0. This resulted in a total of 15 samples with final superficial Si:Al ratios of 1.15, 1.40, 1.65, 1.90 and 2.15.

Microstructural Analysis

Microstructural analysis was performed using an FEI XL-30 FEG-SEM on samples polished using consecutively finer media, prior to final preparation using 1μm diamond paste on cloth. As geopolymers are intrinsically non-conductive, samples were coated using a gold/palladium sputter coater to ensure there was no arching or image instability during image collection. Control samples were prepared using different coating thicknesses, a different coating medium (Osmium) and left uncoated (analysed in a FEG-ESEM with 2 Torr pressure) to ensure microstructural detail was not an artefact of sample preparation.

Compressive Strength and Density

Ultimate compressive strength and Young's modulus were determined using an Instron Universal Testing Machine using a strain rate of 0.60mm/min. Average sample dimensions of

25mm diameter and 51mm height were used in accordance with industry standards of 2:1 height to diameter ratio. Sample surfaces were polished flat and parallel to avoid the requirement of capping. All values presented in the current work are an average of 6 samples with standard deviation as specified. Superficial sample density was measured by averaging calculated density given by the weight of each of the 6 samples divided by their volume prior to compressive strength testing.

XRD Analysis

XRD samples were prepared by grinding samples in a ring mill until the mean particle size (d_{50}) is less than approximately 10μm. XRD diffractograms were collected on a Philips PW 1800 diffractometer with CuK$_\alpha$ radiation generated at 20 mA and 40kV. Typical specimens were step scanned from 5-70° 2θ at 0.02° 2θ steps integrated at the rate of 4.0 s per step.

RESULTS AND DISCUSSION

XRD diffractograms of geopolymers with varied Si:Al ratio are presented in Figures 1-3. The Na- and K-geopolymers synthesised with Si:Al ratio of 1.15 presented in Figures 1 and 2 respectively exhibit multiple intense diffraction peaks correlating to crystalline aluminosilicates. The Na-geopolymer has been identified as faujasite (Zeolite X or Y), however the K-geopolymer cannot be referenced at this stage., though it is probable that it is a form of potassium zeolite. All other geopolymer XRD diffractograms are almost identical, exhibiting an amorphous hump centred at 27-29°2θ. The small peaks indicated by solid circles indicate small amounts of muscovite ($KAl_2(AlSi_3O_{10})(OH)_2$) present in the metakaolin used in synthesis. No zeolite is detected in the XRD diffractogram of NaK-geopolymer (Figure 3), suggesting mixed-alkali interactions interfere with crystallite formation. The peaks associated with muscovite are seen to decrease in relative intensity in NaK- and K-geopolymers (Figures 3 and 2) respectively compared to that seen in Na-geopolymers (Figure 1). The decrease in intensity suggests that the amount of unreacted muscovite present in geopolymers decreases as sodium is replaced by potassium in the activating solutions. The intensity of peaks in NaK and K-geopolymers are similar suggesting similar amounts of muscovite are present in these geopolymers. These trends have been observed in an investigation utilising ^{27}Al MAS-NMR to quantify the amount of metakaolin remaining in geopolymers after curing by comparing the integrated intensity of Al(VI) remaining in hardened samples[13]. The inability to identify differences in the structural characteristics of geopolymers with different Si:Al ratios and alkali composition reinforces that XRD is in general a very poor method for characterising geopolymers. Regardless, XRD is still one of the most commonly used methods for analysis of geopolymer matrices. ^{29}Si MAS-NMR spectra of the 15 samples presented in this work have been investigated in detail previously, demonstrating that acute differences in the molecular structure of geopolymers can be determined using more sensitive techniques[12].

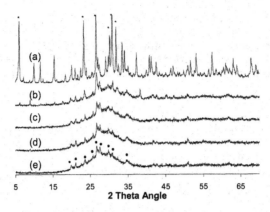

Figure 1 XRD diffractograms of Na-geopolymers with Si:Al ratios of (a) 1.15 , (b) 1.40, (c) 1.65, (d) 1.90 and (e) 2.15 (* indicates intense peaks associated with faujasite and • indicates peaks of muscovite)

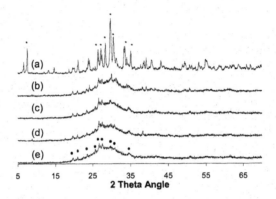

Figure 2 XRD diffractograms of K-geopolymers with Si:Al ratios of (a) 1.15 , (b) 1.40, (c) 1.65, (d) 1.90 and (e) 2.15 (* indicates intense peaks associated with K-Zeolite and • indicates peaks of muscovite)

The results of the compressive strength tests (7-day) on the various geopolymeric binders are summarised in Figure 4. The strength of geopolymers is seen to improve with increasing Si:Al ratio in each of the Na-, NaK- and K- samples. The dramatic improvement in strength in geopolymers from $1.15 \leq$ Si:Al ≤ 1.65 compared to that observed from $1.65 \leq$ Si:Al ≤ 2.15 will be discussed later in this article. Despite previous studies observing that the strength of geopolymers synthesised using potassium containing alkali activating solutions are superior to those synthesised using sodium activating solutions, this study finds that despite there being small observed differences in the strengths of geopolymers synthesised using different alkali, all results fall within one standard deviation from measured strengths for each composition. The

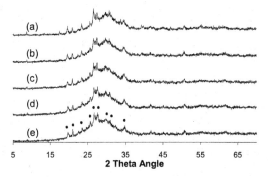

5   15   25   35   45   55   65
2 Theta Angle

Figure 3 XRD diffractograms of NaK-geopolymers with Si:Al ratios of (a) 1.15 , (b) 1.40, (c) 1.65, (d) 1.90 and (e) 2.15 (• indicates peaks of muscovite)

similarity in the ultimate compressive strengths of geopolymers of differing alkali in this study may be reflective of the sample preparation technique, which relies on samples being sealed from the atmosphere at all times prior to testing. All previous studies of geopolymer strength involve curing and storage in environments less than 100% relative humidity, typically 50%[24]. Unsaturated environments create a driving force for water to migrate out of the geopolymeric binder during curing and storage, with K-geopolymers losing more solution than Na-geopolymers. These differences in effective curing conditions could be responsible for the previously reported differences in strength as a result of different alkali cations.

Sample dehydration during curing will limit the extent of geopolymerisation in several ways. Firstly, reducing the amount of water for a given amount of soluble species will increase their concentration, reducing the length of time for gelation and precipitation to initiate. Secondly, less water in the reaction slurry reduces the amount available for hydrolysis reaction during dissolution, which will reduce the extent of reaction. Finally, differences in the energies of hydration of sodium and potassium in aqueous solution will result in geopolymers containing different compositions of alkali to interact differently in reduced humidity environments. In the current study, all geopolymers were sealed from interaction with the environment during all stages of curing and storage prior to compressive strength testing and microstructural analysis as indicated by the negligible weight loss observed in all samples.

The high standard deviations in the value of the ultimate compressive strength of geopolymers at each composition, indicated in Figure 4 by error bars, is reflective of the influence of sample defects such as entrained air. The increases in standard deviation are likely to be due to the increased viscosity of geopolymer slurries with high Si:Al ratios, decreasing the efficiency of vibration for removing bubbles and inconsistencies in the slurries.

The average values of the Young's modulus for each composition are presented in Figure 5, with the standard deviation indicated with error bars. It can be readily observed that Na-geopolymers are nominally 40% stiffer than K-geopolymers. The Young's moduli of NaK-geopolymers are only slightly higher than K-geopolymers, suggesting that the stiffness of the binder is subject to mixed-alkali effects, with a non-linear response to alkali composition, similar to that observed in glass materials. Despite variation in the ultimate compressive strength of geopolymers of the same composition, the standard deviation associated with the Young's modulus therefore, calculated from the linear region of each stress/strain plot is considerably lower. Young's modulus provides a more accurate method for comparison of the mechanical properties of geopolymers in this study.

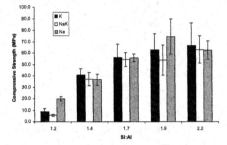

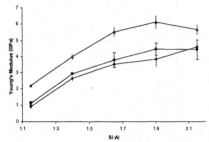

Figure 4 Ultimate compressive strength (7-day) of metakaolin-based geopolymers.

Figure 5 Young's modulus of metakaolin-based geopolymers: (▲) Na-geopolymers, (■) NaK-geopolymers and (♦) K-geopolymers.

A typical plot of stress/strain data collected during compressive strength testing is presented in Figure 6. The plot is characterised by what appears to be an initial period of elastic deformation, typically lasting until approximately 25-30% of the total strain at failure and 18-25% of the ultimate compressive strength is reached. The nature of the deformation in the early stages of compression should be investigated further with load/unload cycling. However, this analysis is not within the scope of this work and is not included. The linear region after elastic deformation obeys Hooke's law and is used to determine the Young's modulus of each sample presented in Figure 5. The characteristics observed in the stress/strain plot in Figure 6 do not vary with Si:Al ratio or alkali composition. The fracturing of geopolymers occurs rapidly without yielding, indicating that geopolymers can be characterised as being a brittle material. The region of plastic deformation is most likely the result of compression within the intrinsically highly porous binder. The stress/strain relationship observed in geopolymers is different from that observed in traditional ordinary Portland cements, which exhibit strain softening beyond the ultimate compressive strength. Although geopolymers exhibit ultimate compressive strengths typical or in excess of traditional cements, their Young's moduli are typically 80% lower.

SEM micrographs of Na-geopolymers exhibit significant change in the characteristic microstructure of the system between Si:Al ratios of 1.40 - 1.65 (Figure 7). Samples with Si:Al < 1.65 exhibit a largely porous microstructure comprised of loosely structured precipitates and unreacted metakaolin. Geopolymers with Si:Al ratio ≥ 1.65 are categorised by a largely homogenous binder containing unreacted particles and micro-pores. The fundamental change in the microstructure of geopolymers as Si:Al ratio is increased above 1.65 is mirrored in the improvement in the ultimate compressive strength of geopolymers in this region (Figure 4). The microstructures of geopolymers with Si:Al ratio ≥ 1.65 are fundamentally identical to the naked eye. It is important that the cracking of samples (Figure 7) is due to dehydration during sample preparation for SEM. Although not shown, the microstructure of geopolymers investigated in the current work does not vary in any qualitative aspect with alkali composition (ie. similar microstructures are observed for NaK- and K-geopolymers).

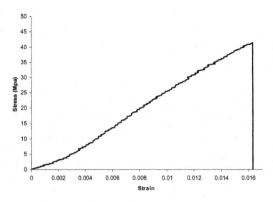

Figure 6 Stress/Strain plot of K-geopolymer with Si:Al of 1.65.

Pore sizes in the order of <5μm are visible in the microstructures of geopolymers with Si:Al ratio ≥ 1.65. The binder at the walls of these pores can be seen to have a layered texture, indicating that they result from particle pullout of soft metakaolin during polishing. The fractional area of the pores caused by particle pullout indicates that the amount of unreacted material in the samples once cured is significant. Unreacted particles can be seen loosely wedged in the structure of geopolymers with Si:Al ratio < 1.65 and appear not to be closely held by the binder. However, it is unclear whether the amount of unreacted particles in geopolymers with Si:Al < 1.65 is less than that observed in samples with Si:Al ≥ 1.65. The extent that metakaolin has been reacted can therefore not be accurately determined from SEM micrographs.

The increase in the apparent homogeneity and density of the microstructure seen in Figure 7 is closely reflected in the large increases in the ultimate compressive strength (Figure 4) and Young's modulus (Figure 5) of the samples as would be expected. The large changes in mechanical properties cannot be predicted by trends in the XRD diffractograms of the geopolymers presented earlier (Figures 1-3), reinforcing the inability of XRD to effectively characterise geopolymers. The superficial densities of Na-, NaK- and K- geopolymers are plotted in Figure 8. The density of geopolymers is seen to increase linearly by only 6% in the range of 1.15 ≤ Si:Al ≤ 2.15; hence the improvement in the mechanical properties must result from the differences in microstructure and not simple correlation between strength and density. A multinuclear NMR structural study of geopolymers identified that the pore solution becomes increasingly constrained as the Si:Al ratio is increased, inferring a decrease in pore size[13]. The weight fraction of water in geopolymers in the current work ranges from 23-35%, representing a significant fraction of the microstructure. The change in the distribution of pore solution from large connected pores to smaller distributed pores within a homogenous microstructure as the Si:Al ratio increases plays a major role in supporting the binder and improving the mechanical properties.

It appears that the microstructure of the geopolymeric binder in the areas surrounding pores created by particle pullout is different than in the bulk. Figure 9(a) shows a SEM micrograph of an Na-geopolymer with Si:Al of 2.15. The voids in the image are created from particle pullout. The microstructure at the interface between the particle voids and the binder is seen to be less dense and continuous. There is no layered pattern at the interface of the particle voids on this scale that would indicate that the binder closely followed the contours of unreacted material. Figure 9(b) shows the porous microstructure of a Na-geopolymer with Si:Al of 2.15 near the site of particle pullout. Spherical precipitates appear at the interface between the binder and the location of the particle, whereas the binder is homogenous only a small distance away from the particle void as seen in Figure 9(a). The structure of the binder at the particle-binder interface is akin to that of geopolymers with Si:Al ratio ≤ 1.65.

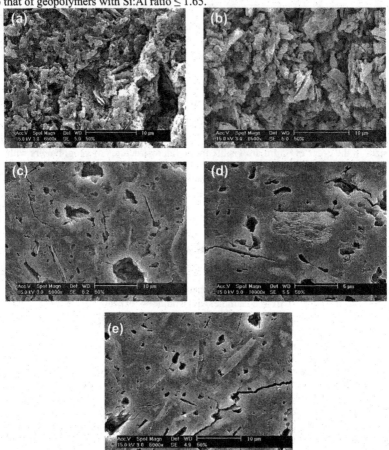

Figure 7 SEM micrographs of Na-Geopolymers: Si:Al ratio of (a) 1.15, (b) 1.40, (c) 1.65, (d) 1.90 and (e) 2.15.

The development of a different microstructures observed by XRD and SEM in geopolymers with $1.15 \leq$ Si:Al ratio ≤ 2.15 must result from effects arising from increases in the concentration of silicon on the activating solution. Furthermore, the change in microstructure from homogeneous to porous at the particle-binder interface must be understood in terms of compositional and chemical differences. The actual dissolution and precipitation processes occurring during geopolymerisation are complex and as yet not fully understood from a fundamental perspective. However, microstructural differences identified in the current work can possibly be understood by considering the basic chemical and transport processes occurring during geopolymerisation.

Geopolymerisation begins when the solid aluminosilicate source and the activating solution are combined. The initial rate of aluminosilicate dissolution is high due to dissolution of fines[25]. Studies of aluminosilicates have shown that early dissolution results in high levels of aluminium being released and the formation of a dealuminated layer on the particle surface[25]. Following this early period, stoichiometric dissolution of aluminium and silicon monomers from the particle occurs. The rates of dissolution have been shown to decrease in the presence of increasing concentration of aluminium and silicon in solution. Therefore, the initial concentration of silicon in the activating solution will largely control initial rates of dissolution and the solution phase composition. Mechanical mixing during the early stages of reaction ensures that soluble aluminium and silicon will be evenly distributed in the solution phase.

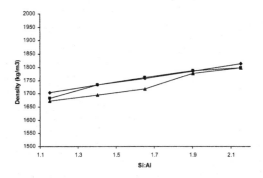

Figure 8 Superficial density of metakaolin-based geopolymers: (▲) Na-geopolymers, (■) NaK-geopolymers and (♦) K-geopolymers.

After mixing ceases, the samples are placed in a static environment. Dissolution will continue, but at a reduced rate (ie. silicon and aluminium will be released as the particle size reduces). Since there is no bulk mixing, a saturated layer of dissolved species will form surrounding particles. The composition of the saturated layer will be controlled by that of the particle[25]. The localised increase of species will create a concentration gradient, which acts as the driving force for mass transport by diffusion. The rate of diffusion will be determined primarily by the size of the concentration gradient and viscosity of the solution phase. Hence, the high viscosity of high Si:Al ratio slurries will reduce the rate of diffusion, which has been linked to a reduced extent of metakaolin reaction[13]. Since the activating solution of geopolymers with Si:Al ratio of 1.15 contains no soluble silicon or aluminium at the time of mixing, the concentrations of aluminium and silicon in the bulk result only from dissolution. Hence, after mixing, the concentrations of soluble species would be similar to that in the saturated particle layer, resulting in minimal driving force. However, where high concentrations of silicon are in the activation solution at the time of initial mixing, the concentration of aluminium in the bulk will be lower than at the saturated layer around the particles inducing a larger concentration gradient.

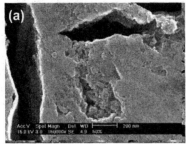

Figure 9 SEM micrographs of Na-geopolymer with Si:Al of 2.15

The concentration of aluminium and silicon in solution has been shown to be well above supersaturation levels[17], however the metastability of concentrated aluminosilicate solutions retards nucleation[26]. A combination of curing temperature, raw material type, particle size, surface area, mixing time and activating solution composition will determine the time at which heterogeneous nucleation from the supersaturated solution will occur. The porous region seen in geopolymers with Si:Al < 1.65 and at the particle/binder interface suggest that the binder forms in regions in the order of 5-10nm, linking in the bulk to form a homogeneous network. The concentration gradient expected from the particle-binder interface to the bulk should be detectable by EDS. Unfortunately, the distance between particles is not sufficient to detect small perturbations in chemical composition at the resolution of EDS using a SEM. Since no concentration gradient is expected in geopolymers with Si:Al ratio of 1.15, a homogeneous composition in the slurry combined with a higher water content is proposed to allow for highly ordered products with a longer range order, as seen in Figures 1-2.

Average chemical compositions of the binder in each geopolymer are shown in Figure 10, formed from the average of 10 spot analyses. The chemical compositions of the binder determined by EDS in geopolymers with Si:Al ratios < 1.65 are similar to expected values when compared to the theoretical superficial Si:Al ratio. If the amount of unreacted material was high, the Si:Al ratio of the binder would be higher than expected, since the Si:Al ratio of metakaolin is 1.15. Hence, only a low level of unreacted metakaolin is predicted from EDS analysis in these geopolymers. The increase of Si:Al ratios in the binder above theoretical values with Si:Al $\geq$ 1.65 reflects the influence of unreacted metakaolin on the binder composition at these compositions. Investigation of the binder composition at higher resolution, such as that available using scanning TEM, should be able to detect any gradient in chemical composition. Nonetheless, the composition of the binder was found to be similar to expected compositions at each spot of analysis, inferring that no large variation in binder composition could be detected.

Low concentrations of silicon in the activator solutions have been shown to inhibit homogeneous binder formation, tending to formation of secondary precipitates on particle surfaces[17], as is also observed in Figure 7a-b. Where the activating solution contains significant concentrations of silicon, homogeneous binders form, however the porous regions surrounding the particles are not predicted. As the saturated layer of aluminium and silicon forms around the static particles at the time of heterogenous nucleation, the composition of region at the particle-binder interface will be analogous to that seen in geopolymers with Si:Al ratio of 1.15, hence a similar microstructure resulting. The similar strengths of geopolymers with Si:Al ratio $\geq$ 1.65, that exhibit a homogenous microstructure, may result from the weak porous regions at the particle-binder interface being the point of fracture initiation.

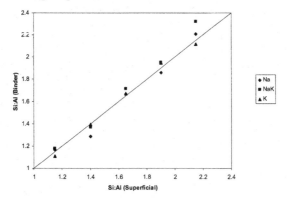

Figure 10 Comparison of theoretical and actual composition of binder in metakaolin-based geopolymers

CONCLUSION

The microstructure of geopolymers has been shown to vary with the superficial Si:Al ratio, regardless of alkali metal. Geopolymers with Si:Al ratio < 1.65 exhibit a highly porous microstructure observed by SEM. Geopolymers with Si:Al ratio ≥ 1.65 are homogenous. The ultimate compressive strengths and Young's moduli of geopolymers with Si:Al ratio ≥ 1.65 are greatly enhanced compared to those of lower Si:Al ratio. Na-geopolymers exhibit higher Young's moduli than NaK- and K-geopolymers, however ultimate compressive strengths are similar in all cases. The similar mechanical properties of geopolymers with homogenous microstructures are proposed to be a result of porous regions of binder at the particle-binder interface. The porous binder is expected to be weak and be the initiation point during fracture.

REFERENCES

[1]J. Davidovits, "Geopolymers - Inorganic polymeric new materials," *Journal of Thermal Analysis*, **37**(8) 1633-1656 (1991).

[2]V.D. Glukhovsky, "Ancient, modern and future concretes," in *Proceedings of the First International Conference on Alkaline Cements and Concretes*. Kiev, Ukraine: VIPOL Stock Company: 1-9. P.V. Krivenko, Ed. (1994)

[3]V.F.F. Barbosa and K.J.D. MacKenzie, "Thermal behaviour of inorganic geopolymers and composites derived from sodium polysialate," *Materials Research Bulletin*, **38**(2) 319-331 (2003).

[4]H. Xu and J.S.J. van Deventer, "The geopolymerisation of alumino-silicate minerals," *International Journal of Mineral Processing*, **59**(3) 247-266 (2000).

[5]A. Palomo and F.P. Glasser, "Chemically-bonded cementitious materials based on metakaolin," *British Ceramic Transactions and Journal*, **91**(4) 107-112 (1992).

[6]G. Engelhardt, D. Hoebbel, M. Tarmak, A. Samoson and E. Lippmaa, "^{29}Si-NMR-Untersuchungen zur Anionenstruktur von kristallinen Tetramethylammonium-alumosilicaten und -alumosilicatlösungen," *Zeitschrift für Anorganische und Allgemeine Chemie*, **484** 22-32 (1982).

[7]H. Xu and J.S.J. van Deventer, "Effect of source materials on geopolymerization," *Industrial & Engineering Chemistry Research*, **42**(8) 1698-1706 (2003).

[8]C.K. Yip and J.S.J. van Deventer, "Microanalysis of calcium silicate hydrate gel formed within a geopolymeric binder," *Journal of Materials Science*, **38**(18) 3851-3860 (2003).

[9]W.K.W. Lee and J.S.J. van Deventer, "Effects of anions on the formation of aluminosilicate gel in geopolymers," *Industrial & Engineering Chemistry Research*, **41**(18) 4550-4558 (2002).

[10]J.G.S. van Jaarsveld, J.S.J. van Deventer and G.C. Lukey, "The effect of composition and temperature on the properties of fly ash- and kaolinite-based geopolymers," *Chemical Engineering Journal*, **89**(1-3) 63-73 (2002).

[11]A. Palomo, M.T. Blanco-Varela, M.L. Granizo, F. Puertas, T. Vazquez and M.W. Grutzeck, "Chemical stability of cementitious materials based on metakaolin," *Cement and Concrete Research*, **29**(7) 997-1004 (1999).

[12]P. Duxson, J.L. Provis, G.C. Lukey, J.S.J. van Deventer and F. Separovic, "^{29}Si MAS NMR Investigation of Metakaolin-Based Geopolymers. I -Determination of Silicon Distribution," *Submitted - Journal of Physical Chemistry B* (2004).

[13]P. Duxson, G.C. Lukey, F. Separovic and J.S.J. van Deventer, "Multinuclear NMR structural studies of metakaolin-based geopolymers," *Submitted - Solid-State Nuclear Magnetic Resonance* (2004).

[14]M. Rowles and B. O'Connor, "Chemical optimisation of the compressive strength of aluminosilicate geopolymers synthesised by sodium silicate activation of metakaolinite," *Journal of Materials Chemistry*, **13**(5) 1161-1165 (2003).

[15]P.S. Singh, M. Trigg, I. Burgar and T. Bastow, "^{27}Al and ^{29}Si MAS-NMR Study of Aluminosilicate Polymer Formation at Room Temperature," in *Geopolymers 2002. Turn Potential into Profit*. Melbourne, CD-ROM Proceedings: University of Melbourne. G.C. Lukey, Ed. (2002)

[16]V.F.F. Barbosa, K.J.D. MacKenzie and C. Thaumaturgo, "Synthesis and characterisation of materials based on inorganic polymers of alumina and silica: sodium polysialate polymers," *International Journal of Inorganic Materials*, **2**(4) 309-317 (2000).

[17]W.K.W. Lee and J.S.J. van Deventer, "Structural reorganisation of class F fly ash in alkaline silicate solutions," *Colloids and Surfaces A - Physicochemical and Engineering Aspects*, **211**(1) 49-66 (2002).

[18]H. Xu and J.S.J. van Deventer, "Microstructural characterisation of geopolymers synthesised from kaolinite/stilbite mixtures using XRD, MAS-NMR, SEM/EDX, TEM/EDX, and HREM," *Cement and Concrete Research*, **32**(11) 1705-1716 (2002).

[19]W.M. Kriven, J.L. Bell and M. Gordon, "Microstructure and microchemistry of fully-reacted geopolymers and geopolymer matrix composites," *Ceramic Transactions*, **153** 227-250 (2003).

[20]W.M. Kriven and J.L. Bell, "Effect of alkali choice on geopolymer properties," *Ceramic Engineering and Science Proceedings*, **25**(3-4) In press (2004).

[21]C.K. Yip, G.C. Lukey and J.S.J. van Deventer, "Microstructure and properties of slag/metakaolinite geopolymierc materials," in *28th Conference on Our World in Concrete and Structures*. Singapore. (2003)

[22]H. Rahier, J.F. Denayer and B. van Mele, "Low-temperature synthesized aluminosilicate glasses. Part IV. Modulated DSC study on the effect of particle size of metakaolinite on the production of inorganic polymer glasses," *Journal of Materials Science*, **38**(14) 3131-3136 (2003).

[23]J.P. Hos, P.G. McCormick and L.T. Byrne, "Investigation of a synthetic aluminosilicate inorganic polymer," *Journal of Materials Science*, **37**(11) 2311-2316 (2002).

[24]W.K.W. Lee and J.S.J. van Deventer, "The use of infrared spectroscopy to study geopolymerization of heterogeneous amorphous aluminosilicates," *Langmuir*, **19**(21) 8726-8734 (2003).

[25]E.H. Oelkers, J. Schott and J.L. Devidal, "The effect of aluminum, pH, and chemical affinity on the rates of aluminosilicate dissolution reactions," *Geochimica et Cosmochimica Acta*, **58**(9) 2011-2024 (1994).

[26]J. Faimon, "Oscillatory silicon and aluminum aqueous concentrations during experimental aluminosilicate weathering," *Geochimica et Cosmochimica Acta*, **60**(15) 2901-2907 (1996).

ACKNOWLEDGEMENT
Financial support is gratefully acknowledged from: the Australian Research Council (ARC), the Particulate Fluids Processing Centre (PFPC), a Special Research Centre of the Australian Research Council, and the United States Air Force Office of Scientific Research (AFOSR), under STTR Grant number F49620-02 C-010 in association with The University of Illinois at Urbana-Champaign and Siloxo Pty Ltd, Melbourne, Australia.

THE EFFECT OF HEAT ON GEOPOLYMERS MADE USING FLY ASH AND METAKAOLINITE

Dan S Perera, Eric R Vance, David J Cassidy, Mark G Blackford, John V Hanna, Rachael L Trautman
Australian Nuclear Science and Technology Organisation
Private Mail Bag 1, Menai, NSW 2234, Australia
Catherine L Nicholson
Industrial Research Ltd.
P Box 31-310, Lower Hutt, New Zealand

ABSTRACT
Three geopolymers were investigated which have potential for immobilization of radionuclides such as Cs and Sr, which are two of the most difficult radionuclides to immobilize. In geopolymers for immobilizing low/intermediate level radioactive waste an intrinsic difficulty is radiolytic gas generation from the occluded H_2O/OH^- in the structure. The effect of heat on these geopolymers was studied with the aim of removing these species and so avoiding this gas generation.
Two geopolymers were made by dissolving metakaolinite or fly ash in concentrated $NaOH/Na_2SiO_3$ solutions. Another commercial metakaolinite geopolymer was also investigated. One weight % Cs or Sr was added to the geopolymers. Major loss of water on heating of the cured geopolymers occurs from ambient to ~300^{0}C as deduced by differential thermal analysis/thermogravimetric analysis. The geopolymers were progressively heated up to 1200^{0}C in air and the phase changes were studied by X-ray diffraction (XRD) analysis and scanning electron microscopy. After heating at 500^{0}C the XRD analysis showed the same phases as at ambient for each geopolymer. Transmission electron microscopy also did not show any significant differences. The Cs and Sr were present in the amorphous phase. After heating above 900^{0}C, Na or K feldspar was present in all and the presence of other phases was dependent on the type of geopolymer. The porosity increased after heating between 500-900^{0}C and then decreased due to sintering at up to 1200^{0}C. The ^{133}Cs MAS NMR spectra for the Cs-containing fly ash geopolymer at ambient and at 500^{0}C showed large line widths which indicates that the Cs is held mostly in the geopolymer network rather than in the interstitial water. The 500^{0}C spectrum was slightly wider than the ambient which indicates more Cs is held in the geopolymer network at the higher temperature. This investigation shows that geopolymers heated to 500^{0}C remain as promising candidates for low/intermediate level radioactive waste immobilization after the elimination of most of the interstitial water and hydroxyl ions.

INTRODUCTION
Inorganic polymers formed from naturally occurring aluminosilicates have been termed geopolymers by Davidovits [1]. Various sources of Si and Al, generally in reactive glassy or fine ground forms, are added to concentrated alkali solutions for dissolution and subsequent polymerisation to take place. Typical precursors used are fly ash, ground blast furnace slags, metakaolinite made by heating kaolinite at ~ 750^{0}C for 6-24 h, or other sources of Si and Al. The alkali solutions are typically a mixture of hydroxide (e.g. NaOH, KOH), or silicate (Na_2SiO_3, K_2SiO_3). The solution dissolves Si and Al ions from the precursor to form $Si(OH)_4$ and $Al(OH)_4$ monomers in solution [2]. In this structure Al is tetrahedrally coordinated to O, hence the monomer has a negative charge. The non-network alkali cations balance this charge. The OH^- ions of neighbouring molecules condense to form an oxygen bond between metal atoms and release a

molecule of water. Under the application of low heat (20-90^{0}C) the material polymerizes to form a rigid polymer containing interstitial water. The polymers consist of amorphous to semi-crystalline two or three dimensional aluminosilicate networks, dependent on the Si to Al ratio [1]. Their physical behaviour is similar to that of Portland cement and they have been considered as a possible improvement on cement in respect of compressive strength, resistance to fire, heat and acidity, and as a medium for the encapsulation of hazardous or low/intermediate level radioactive waste (LLW/ILW) [3-6]. Although they have been used in several applications their widespread use is restricted due to lack of long term durability studies, detailed scientific understanding and lack of reproducibility of raw materials.

Cs and Sr are two of the most difficult radionuclides to immobilize because of their solubility in water and are therefore key elements to study in assessing geopolymers as matrices for immobilization of radioactive wastes. In geopolymers for immobilizing LLW/ILW an intrinsic difficulty is radiolytic gas generation from the occluded H_2O/OH^- in the structure. In this study geopolymer samples were prepared with either Cs or Sr from fly ash and metakaolinite precursors. For comparison a commercial metakaolinite geopolymer was also used. The materials were heated from ambient to 1200^{0}C after curing to study microstructural changes..

EXPERIMENTAL

The batch compositions of the fly ash geopolymer (FA) consisted of 68.8 wt% NZ fly ash, 3.2 wt% NaOH, 13.7 wt% Na_2SiO_3 and 14.3 wt% demineralized water and they were made as described previously [6]. The method of making the geopolymers, including that of the commercial metakaolinite geopolymer (Davya 60, supplied by CORDI-Geopolymère, France) and referred to as MK geopolymer, have also been described in detail [6, 7]. A batch of MK geopolymer was also made with ~ 1 wt% Cs added as the oxide. Another metakaolinite geopolymer (MK1) was made by adding 38.5 wt% commercial metakaolinite to a solution consisting of 6.3 wt% NaOH dissolved in 9.4 wt% demineralized water and 44.3 wt% Na_2SiO_3. One wt% Cs or Sr was added as the hydroxide or the nitrate to the FA and MK1 geopolymers. All the geopolymers were cured for 2 h at ambient followed by 18-24 h at 80-90^{0}C and tested after a further 6 days.

To study the effect of heating on the microstructure and loss of water and other species, the cured pastes were heated from 500-1200^{0}C for 3 h in an electric furnace with heating and cooling rates of 5^{0}C/min. Differential thermal analysis/thermogravimetric analysis (DTA/TGA: Setaram Tag24, simultaneous Differential Thermal Thermogravimetric Analyzer, France) was also carried out on FA and MK geopolymers. The sample weight was ~ 85 mg and was placed in a Pt crucible with Al_2O_3 as the reference sample. The analyses were carried out in air with a heating rate of 5^{0}C/min in flowing air (25 mL/min) to 850^0 or 1050^{0}C .

The density and porosity of each of the geopolymers were determined according to the Australian Standard [8] by evacuating under vacuum and introducing water to saturate the pores. The time of saturation and the immersion in water was kept to less than 15 min to inhibit reaction with water.

All samples were analysed by X-ray diffraction (XRD: Model D500, Siemens, Karlsruhe, Germany) using CoKα radiation on crushed portions of material. Selected samples were cross sectioned, mounted in epoxy resin and polished to a 0.25 μm diamond finish and examined by scanning electron microscopy (SEM: Model 6400, JEOL, Tokyo, Japan) operated at 15 kV and fitted with an X-ray microanalysis system (EDS: Model: Voyager IV, Tracor Northern, Middleton, WI, USA).

Transmission electron microscopy (TEM, JEOL model JEM 2000FXII) was carried out at 200 kV on FA at ambient and at 500^{0}C. The TEM was equipped with a LINK ISIS X-ray microanalysis system (Oxford Instruments, UK). TEM specimens were prepared by suspending a small amount of powdered material in AR-grade ethanol and depositing several drops by pipette onto a holey carbon coated copper TEM grid. Energy dispersive X-ray spectroscopy (EDS) was employed to determine the chemical composition of the various phases identified in the samples. Selected area electron diffraction (SAED) patterns were observed to determine whether areas of the sample were crystalline or amorphous.

Magic angle spinning nuclear magnetic resonance (MAS NMR) was used to study the Cs speciation in FA at ambient and at 500^0C. The experimental details of the method have been reported previously [6].

RESULTS AND DISCUSSION

Microstructure

The values of density and porosity are listed along with XRD analyses of the samples in Table 1. The open porosities of all the geopolymers increase and then decrease with increase of heat-treatment temperature. The most likely explanation is that the increase in porosity is due to the removal of water and breaking of silanol bonds at 500^0C, causing the opening of pores. The porosity decrease from 800-1200^{0}C is attributed to sintering in the presence of a liquid phase and partial or complete melting at the highest temperature. It is quite feasible to envisage the presence of a liquid phase at 800^{0}C for a system consisting of Na_2O-CaO-Al_2O_3-SiO_2, when the lowest eutectic temperature for the Na_2O-CaO-SiO_2 alone is 725^{0}C [9]. The FA and MK1 geopolymers had melted and formed a globule at 1200^{0}C and had almost no porosity, indicating that they were glassy which was also confirmed by XRD analysis (Table 1) on the cooled material. The MK geopolymer also progressively and significantly decreased in porosity after heating from 800^0 to 1200^{0}C. It showed a shiny surface, probably due to partial melting after heating to 1200^{0}C.

From XRD, the FA and MK1 geopolymers after heating at 900^{0}C exhibited Na or K feldspars. The FA geopolymer showed the presence of gehlinite at 900^{0}C and MK geopolymer contained akermanite at 800^{0}C, because it contains substantial amounts of Mg [6].

The SEM image for MK1 heated to 500^{0}C shows (Figure 1(left)) undissolved kaolinite as unconverted material in the commercial metakaolinite. The EDS analysis of the kaolinite showed an atomic ratio of 1:1 for Al to Si as expected. The presence of the kaolinite was also confirmed by XRD (Table 1). The microstructure of the 500^{0}C heated material was similar to that of the unheated geopolymer. The SEM image of MK1 heated to 900^{0}C shows (Figure 1(center)) the nepheline syenite phase as the dark grey phase. The fine white phase showed the presence of Si, Al, Ti, Na, O, which is probably the undissolved kaolinite (also deduced by XRD – see Table 1) but present to a lesser extent than at 500^{0}C. The black regions are pores. The SEM image of MK1 heated to 1200^{0}C shows (Figure 1(right)) the glassy matrix and a small amount of porosity (it showed 0.5% open porosity, Table 1). The large pore (partly shown in Figure 1 (right)) is probably a gas bubble from minor foaming. The EDS analysis of the matrix phase showed the presence of Cs cations varying from 1.30 to 1.6 wt% for all heated MK1 geopolymers. The values were higher than the 1% Cs added because none could enter unreacted materials or some of the phases that crystallized on heating may also be due to water loss.

The TEM images of the amorphous regions which was the major phase of the FA geopolymer, showed the presence of Cs and Sr in the Cs-containing and Sr-containing geopolymers respectively. The amount of Cs and Sr in the amorphous phases varied from 2 to 4 wt%. In the Sr-containing sample there were a few unreacted Sr particles. The TEM micrographs also showed undissolved fly ash and crystalline phases. The latter consisted of mullite and quartz which were present in the original fly ash. The ambient and 500^{0}C heated geopolymers were similar, except the Cs-containing geopolymer (500^{0}C) showed a few particles with needle-like crystals of Al and Si oxide within the amorphous phase (Figure 2). Because of the small dimensions of the TEM specimens, it is not clear whether they had grown from the amorphous phase due to heating or whether they were present

Table I. Bulk density, open porosity and XRD analysis of geopolymers

Temp, ^{0}C	BD g/cm^3	OP %	XRD analysis
FA GP			
25	1.71	27.2	Am (M), Q, calcite, mullite, calcium silicate
500	1.62	38.0	Am (M), Q, calcite, mullite, calcium silicate
900	1.76	33.8	Am, Q (M), mullite, gehlinite, NS, jadeite
1100	1.63	6.1	Am,(M), gehlinite, NS
1200	2.82	0.0	Am
MK1 GP			
RT	1.24	44.3	Am (M), kaolinite
500	1.29	45.0	Am (M), kaolinite
900	2.30	1.5	Am, NS (M), kaolinite
1200	1.91	0.5	Am
MK GP			
RT	1.67	25.8	Am (M), Q, calcite, Ca/Al/Si/Mg hydroxide
500	1.55	60.5	Am (M), Q, calcite, Ca/Al/Mg silicate
800	1.33	51.5	Am, Q, calcite, akermanite (M)
1000	2.27	11.1	Q (trace), akermanite (M), leucite
1200	2.12	3.2	Leucite (M), akermanite, calcium silicates

Note 1: GP = geopolymer; BD = bulk density; OP = open porosity; Am = amorphous; Q = quartz; NS = nepheline syenite; M = major, all the other phases are minor.
Note 2: Calcite: $CaCO_3$; mullite: $3Al_2O_3.2SiO_2$; nepheline syenite: $Na_2O.Al_2O_3.2SiO_2$; jadeite: $Na_2O.Al_2O_3.4SiO_2$; gehlinite: $2CaO.Al_2O_3.SiO_2$; akermanite: $2CaO.MgO.2SiO_2$; leucite: $K_2O.Al_2O_3.4SiO_2$.
Note 3: Amorphous phase deduced from a very broad diffuse peak centred at a d-spacing of ~0.32 nm.

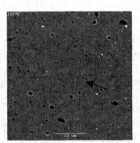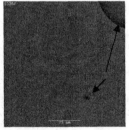

Figure 1. Backscattered SEM images of FA geopolymer heated to (left) 500^{0}C, showing unreacted kaolinite (light grey phase); (center) 900^{0}C, showing the nepheline syenite phase (darker phase, arrow); unreacted kaolinite (white areas) and pores (black areas) and (right) 1200^{0}C, showing some porosity (arrows).

before heating. The Al:Si weight ratio of the crystalline phase was ~ 4:1, which is close to that expected for mullite (3:1) within experimental error. Hence, it is most likely mullite which was present in the fly ash precursor. The amorphous areas produced broad diffuse SAED ring patterns.

DTA/TGA

The FA geopolymer yielded two endothermic peaks (Figure 3) at ~ 80 ^{0}C and 130 ^{0}C compared with the single broad peak at ~ 90^{0}C in the MK geopolymer (Figure 4). The two peaks may result from the fine porous structure usually associated with fly ash, with water from the surface of the sample released mostly at lower temperature, followed by loss of water retained by the pores in the amorphous geopolymer matrix mostly at the higher temperature. The major weight loss of interstitial water occurs between ambient and 300^{0}C. The weight loss in both FA and MK geopolymers between 500-600^{0}C was possibly due to decomposition of calcite which was present in both samples as shown by XRD analysis (Table 1). Above 600^{0}C there is further weight loss, albeit less, probably due to breakdown of silanol bonds and release of hydroxyl ions. There is a small endothermic peak at ~950^{0}C and a larger peak at 1017^{0}C for the FA geopolymer (Figure 3) which indicates localized melting, as discussed above.

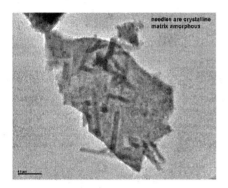

Figure 2. TEM image of the Cs-containing FA geopolymer heated to 500^{0}C, showing needle-like crystals in an amorphous matrix.

Cs speciation

The [133]Cs MAS NMR spectra for the Cs-containing FA geopolymer at ambient and at 500 ^{0}C are shown in Figure 5. All the spectra exhibit markedly increased line widths in comparison to the MK geopolymer [10]. This indicates that the Cs is held more in the geopolymer network than in the interstitial water. If Cs was mainly in the interstitial water then the resonance peaks would be sharper. The high temperature spectrum is slightly wider than the ambient which may indicate more Cs is held in the geopolymer network at the higher temperature. Unpublished leach tests carried out on the finely ground FA geopolymer by digesting with water for 7 days at 90^{0}C (Product Consistency Test- PCT) indicated the presence of Cs and Sr in the water in amounts ~ 100 times less than those for cement. The preliminary data indicate that the release from heated geopolymers in PCT tests [11] are comparable to those of Environmental Assessment glass.

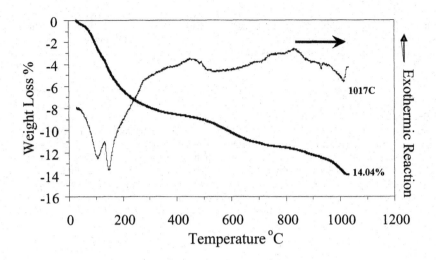

Figure 3. DTA/TGA of FA geopolymer.

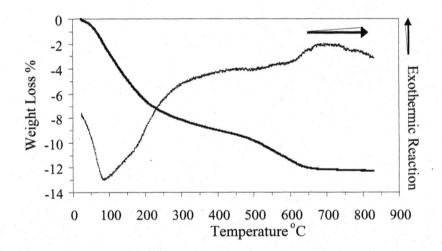

Figure 4. DTA/TGA of MK geopolymer.

CONCLUSIONS

Major loss in water on heating occurs from ambient to ~300°C as deduced by DTA/TGA analysis. At 500°C the XRD analyses showed the same phases as at ambient for all the geopolymers

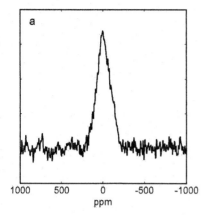
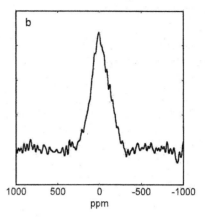

| 1000 | 500 | 0 | -500 | -1000 | 1000 | 500 | 0 | -500 | -1000 |

ppm   ppm

Figure 5. ^{133}Cs NMR spectra of FA geopolymer (a) at ambient (peak at 26.1 ppm) and (b) at 500^{0}C (peak at 14.9 ppm).

investigated. The major phase was amorphous on the 50-100 nm scale from TEM. TEM also did not show any significant differences in the distribution of the amorphous phase (and no sign of incipient crystallization on the 50-100 nm scale) and the amount of Sr and Cs in the amorphous phase at ambient and after heating at 500^{0}C for the FA geopolymer. Above 900^{0}C Na or K feldspar was present in each sample and the presence of other phases was dependent on the type of geopolymer. The porosity increased after heating between 500-900^{0}C and then decreased due to sintering up to 1200^{0}C. The ^{133}Cs MAS NMR spectra for the Cs-containing fly ash geopolymer at ambient and after heating at 500^{0}C showed broad peaks, indicating that the Cs is held mostly in the geopolymer network. The spectrum after heating to 500^{0}C was slightly wider than at ambient which indicates that more Cs is held in the geopolymer network after heating at the higher temperature. This part of theinvestigation of a wider study on geopolymers shows that geopolymers heated to 500^{0}C are promising candidates for low/intermediate level radioactive waste immobilization after the elimination of most of the interstitial water.

ACKNOWLEDGMENTS
The authors thank Ted Roach and Ross Fletcher for technical assistance.

REFERENCES

[1] J. Davidovits, "Geopolymers – Inorganic Polymeric New materials," Journal of Thermal. Analysis., 37 [8] 1633-56 (1991).
[2] P. G. McCormick and J. T. Gourley, "Inorganic Polymers – A new Material for the New Millenium," Materials Australia, 23 16-18 (2000).
[3] J. Davidovits, " Geopolymers: Man-Made Rock Geosynthesis and the Resulting Development of very Early High Strength Cement," Journal Materials Education, 16 [12] 91-139 (1994).
[4] J. Davidovits,"Chemistry of Geopolymeric Systems, Terminology," Geopolymere '99, Geopolymer International Conference, Proceedings, 30 June – 2 July, 1999, pp. 9-39, Saint-Quentin,

France. Edited by J. Davidovits, R. Davidovits and C. James, Institute Geopolymere, Saint Quentin, France (1999).

[5] A. Allahverdi and F. Skvara, "Nitric Acid Attack on Hardened Paste of Geopolymeric Cements," Ceramics-Silikatay, **45** [3] 81-8 (2001).

[6] D. S. Perera, E. R. Vance, Z. Aly, K. S. Finnie, J. V. Hanna, C. L. Nicholson, R. L. Trautman, and M. W. A. Stewart, "Characterisation of Geopolymers for the Immobilisation of Intermediate Level Waste," Proceedings of ICEM'03, Sepetember 21-25, 2003, Oxford, England, Laser Options Inc., Tucson, USA, 2004, CD, paper no. 4589.

[7] D. S. Perera, C. L. Nicholson, M. G. Blackford, R. A. Fletcher and R. L. Trautman, "Geopolymer Made Using New Zealand Fly Ash," Journal of the Ceramic Society of Japan, Supplement 112-1, **112** [5] S108-S111 (2004).

[8] Australian Standard AS 1774.5-2001, "The determination of density, porosity and water absorption", Standards Australia (2001).

[9] "Phase Diagram for Ceramists", Edited by E. M. Levin, C. R. Robbins and H. F. Mc Murdee, P. 175, American Ceramic Society, Westerville, Ohio, USA, 1964.

[10] D. S. Perera, M. G. Blackford, E. R. Vance, J. V. Hanna, K. S. Finnie, and C. L. Nicholson, "Geopolymers for the Immobilization of Radioactive Waste", Presented at the MRS Spring Meeting April 11-15, 2004, San Francisco, to be published.

[11] "Determining Chemical Durability of Nuclear Hazardous and Mixed Waste Glasses: The Product Consistency Test (PCT), ASTM Designation: C 1285-97.

COMPARISON OF NATURALLY AND SYNTHETICALLY-DERIVED, POTASSIUM-BASED GEOPOLYMERS

M. Gordon, J. L. Bell and W. M. Kriven
Department of Materials Science and Engineering,
University of Illinois at Urbana-Champaign,
Urbana, IL, 61801, USA

ABSTRACT
    A potassium cross-linked geopolymer made from synthetic metakaolin (SynMK) was prepared and contrasted to a similar geopolymer derived from natural metakaolin (NatMK). X-ray diffraction (XRD) showed that the SynMK was a high purity analog of NatMK that exhibited no crystallinity and had a diffraction "halo" similar to that of the NatMK at 23° two-theta. The SynMK had a surface area of 166 m^2/g and reacted rapidly with a potassium silicate solution to form a synthetic geopolymer. XRD of the resulting geopolymers, based on NatMK and SynMK, showed that the diffraction "halo" of the starting metakaolins at 23° two-theta shifted to 28° two-theta after reaction with alkali-silicate solutions. SEM micrographs of the geopolymer based on SynMK revealed a uniform and dense microstructure with no unreacted aluminosilicate particles present. Bright field TEM of the synthetic geopolymer showed a two phase microstructure, similar to naturally-based geopolymers described in literature.

INTRODUCTION
    Geopolymers, also called polysialates, are a class of amorphous aluminosilicate materials formed near ambient temperatures. Chemically, geopolymers consist of cross-lined units of AlO_4^- and SiO_4 tetrahedra, where charge-balancing cations are provided by alkali metal cations such as Li+, Na+, K+, and Cs+. Geopolymers are formed by reacting aluminosilicate materials with highly caustic, aqueous alkaline (MOH) silicate solutions, where M= alkali cation. During the synthesis of geopolymers, dissolution of the aluminosilicate precursor takes place first, followed by the cross-linking and formation of the geopolymer structure. The process of "geopolymerization" and the study of the subsequent geopolymers has been the subject of numerous studies.[1-26] Applications of geopolymers have included ceramic matrix composites,[1-6] waste encapsulation[7-9], and alternative cements.[10-19,24]
    With the exception of Hos et al.[26], the aluminosilicate precursors for the fabrication of monolithic geopolymers have been limited to naturally derived sources and industrial waste products, e.g. fly ash, slag, raw minerals and calcined kaolin (metakaolin). In their work, Hos et al. created $Al_2O_3 \bullet 2\ SiO_2$ coupons, which were heated to $1550^\circ C$, melted with an oxy-acetylene torch, and rapidly cooled. The resulting glass, which was milled, had a surface area $15 - 16$ m^2/g, and mean particle size of 3.2 μm. X-ray diffraction showed crystalline peaks of mullite and aluminum silicate hydroxide present in the aluminosilicate powder.
    In order to form a homogeneous, pure and amorphous aluminosilicate starting material the organic steric entrapment method of processing was chosen for this study. The steric entrapment method, similar to the Pechini process[27], has been used to create a wide variety of ceramic oxides including, silica, alumina, mullite, yttrium aluminum garnet (YAG), leucite, xenotime, nickel aluminate, barium titanate, etc.[28-37] During the organic steric entrapment method, metallic salts and a soluble polymer, typically polyvinyl alcohol, are dissolved into solution together and mixed. After mixing, the solute, (water or alcohol) is removed by drying

and the resulting powder is calcined. During calcination, the anionic components of the salts are sublimed and the cation components oxidize. The resulting powders are chemically homogenous, and depending on composition and calcination temperature, may have surface areas in excess of 100 m^2/g.

EXPERIMENTAL PROCEDURES

Synthetic metakaolin (SynMK) was prepared by the organic, steric entrapment method. A 5 wt % solution of Celvol Polyvinyl Alcohol, PVA (Celanese Chemicals, Dallas, Texas) was mixed using deionized water and allowed to stir on a hot plate overnight under warm heat. A second solution, 50 w % aluminum nitrate nonahydrate, of 98 % purity (Fisher Scientific, Pittsburgh, PA) was also prepared using deionized water, and allowed to stir overnight. While operating under a fume hood, the aluminum nitrate nonahydrate solution was poured into a 1.5 L Visions® lithium aluminosilicate piece of glassware (Corning, Corning, NY), while being stirred and heated. Colloidal silica, Ludox SK, 24.5 w% (Grace Division, Columbia, Maryland) was then added to the aluminum nonahydrate solution. The two solutions were allowed to mix for 1 hour before the 5 wt % solution of PVA was added. The final molar ratio of the mixed solution was $3 Al_2O_3 \cdot 4\ SiO_2 \cdot 7$ PVA. Infrared heating lamps were set above the mixture and left operating overnight, while the mixture was simultaneously stirred and dried by a hot/stir plate. After sufficient drying (approximately 12 hours) the solution became a uniform yellow cake, which was pulverized using an alumina mortar and pestle.

The resulting powder was then calcined at 800°C for 1 hour at 5°C / min in a 650 Isotemp Furnace (Fisher Scientific). After calcination, the powder color changed to white. The expected composition of the powder was $Al_2O_3 \cdot 1.3\ SiO_2$. The powder was milled with isopropyl alcohol in an attrition mill (Union Process, Akron, Ohio) using 5 mm yttria stabilized zirconia (3Y-TZP) media. The slurry was dried and baked at 800°C for 5 hours at 5°C / min to remove all organics from the surface of the SynMK. (It had been found previously that trace alcohol on the surface of SynMK prevents reaction with alkali-silicate solutions). The SynMK was then sieved through a 100 μm mesh sieve.

An alkali-silicate solution was mixed using potassium hydroxide pellets (Fisher Scientific), Cab-o-Sil silica fume 99.98 % purity (Cabot, Tuscola, IL) and de-ionized water and allowed to stir overnight. The molar ratio of the alkali-silicate solution was $K_2O \cdot 2\ SiO_2 \cdot 10$ H_2O. Using a polypropylene container, SynMK was added incrementally to an alkali-silicate solution while mixing with a high shear mixer to form a geopolymer gel. During mixing, the alkali-silicate solution was cooled with an ice water bath to prevent rapid setting of the geopolymer. After mixing for 15 minutes, the resulting gel was poured into chemically resistant plastic tubing, sealed using rubber stoppers, and allowed to cure at 40°C for 24 hours. The hardened, potassium-based, geopolymer was removed from the tubing. The resulting geopolymer had a superficial composition of $0.8\ K_2O \cdot Al_2O_3 \cdot 3\ SiO_2$.

For comparison purposes between natural metakaolin (NatMK) and SynMK, Hydrite PXN kaolin (Imerys, Dry Branch, Georgia) was calcined at 700°C for 1 hour at 5°C / min in a 51984 Model furnace (Lindberg). The NatMK was mixed with a second alkali-silicate solution which had been mixed previously with a ratio of $K_2O \cdot 2.1\ SiO_2 \cdot 10.6\ H_2O$. The same mass ratios of metakaolin to alkali-silicate solution were used as had been done with the SynMK. No ice bath was required for mixing the NatMK with alkali-silicate solution. The resulting geopolymer was cured and prepared in the same manner as the synthetically derived geopolymer.

Powders of SynMK, NatMK, and geopolymer were sieved through a 45 µm mesh and characterized using a Rigaku D/Max-b (Phillips) x-ray diffractometer. X-ray diffraction (XRD) was done using a copper target, $\lambda = 1.54056$ at 45 kV and 20 mA. Diffraction scans were performed from $5 - 75°$ two-theta at $0.5° /$ min with a step size of $0.02°$. SynMK and NatMK, were degassed and their densities measured using an Accupyc 1330 Pycnometer (Micromeritics, Norcross, Georgia). Particle size analysis was performed on the powders using a CAPA-700 Particle Size Analyzer (Horbia, Irvine, CA,) with sodium pyrophosphate as a dispersing agent. Surface area analysis of the SynMK was performed at Micromeritics Instrument Corporation, Georgia, USA.

Samples of SynMK, NatMK were sputtered coated with a gold-palladium alloy using a K575 Sputter Coater (Emitech, Houston, Texas). Samples of geopolymer made from SynMK and NatMK were fractured with a razor blade and the fracture surfaces of the geopolymers were sputtered coated. Scanning electron microscopy (SEM) was performed on the samples with a S-4700 SEM (Hitachi).

To prepare a geopolymer sample for transmission electron spectroscopy, a 2 mm thick disc was sectioned from a sample of geopolymer based on SynMK using a 4" diamond saw (Buehler). Faces of the disc were ground parallel using silicon carbide sand paper, before being polished with Emery paper. A micro-section of the geopolymer disc was removed and attached to a copper TEM grid using a Dual-Beam DB-235 focused ion beam and scanning electron microscope (FEI, Hillsboro, Oregon). The micro-section was viewed with 2010 LaB$_6$ TEM in bright field mode at 200kV (JEOL).

RESULTS

The results of the particle size and surface area analysis of the NatMK and SynMK are given in Table 1. NatMK had an average particle size half that of the SynMK, despite milling of the latter. Figure 1 and Fig. 2 show the particle size distribution of the NatMK and SynMK respectively. The SynMK had an unusually large surface area, an order of magnitude higher than that of the NatMK used in the study, or of any naturally occurring metakaolin.

Table 1: Characteristics of NatMK and SynMK powders

	NatMK	SynMK
Density (g/cm^3)	2.54	2.59
Surface Area (m^2/g)	18 ^{6}	166
Median Dia. (µm)	0.79	1.62
Stand. Dev. (µm)	3.13	2.71

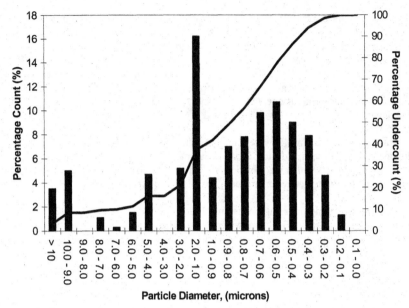

Fig. 1: Particle size distribution of NatMK

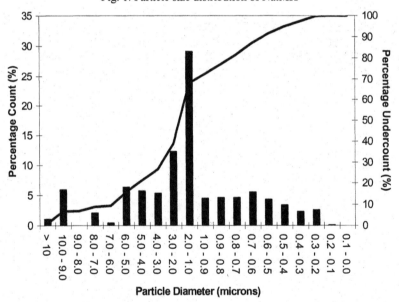

Fig. 2: Particle size distribution of SynMK

XRD of the NatMK, SynMK and the respective geopolymers is given below in Fig. 3. Both the NatMK and the SynMK exhibited a diffraction "halo," at 23° two-theta. The crystalline peaks in the NatMK and its resulting geopolymer are from impurities of anatase, TiO_2. The intensity of the antase peaks decreased during geopolymerization. Minor diffraction may have been present in the SynMK at 68° two-theta, but no distinct crystalline phase existed in the SynMK or its corresponding geopolymer. Geopolymers made from NatMK and SynMK both showed a two-theta shift in the amorphous "halo" in metakaolin from 23° to 28° two-theta. This "halo" shift is similar to what Barbosa et al. observed when metakaolin in their study was reacted with sodium silicate solutions to form geopolymers.[16]

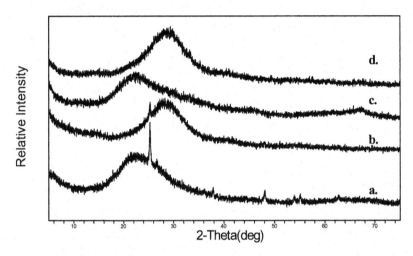

Fig. 3: XRD plots of NatMK (a), geopolymer derived from NatMK (b), SynMK (c) and geopolymer derived from SynMK (d)

SEM analysis of the NatMK and SynMK showed that both powders had sheet-like morphologies. Geopolymers made from the NatMK and SynMK, however, were noticeably different. Unreacted metakaolin sheets were visible in the naturally derived geopolymer. Elongated voids, most likely created when unreacted sheets of metakaolin pull loose from the bulk geopolymer, were also present in the naturally derived geopolymer. The synthetic geopolymer showed no unreacted sheet-like particles and appeared to have a denser microstructure than the geopolymer derived from NatMK.

Fig. 4: SEM micrograph of NatMK, showing sheet-like morphology.

Fig. 5: SEM micrograph of SynMK.

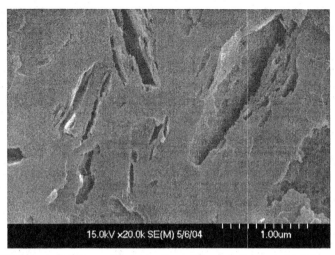

Fig.6: SEM micrograph from a fracture surface of the geopolymer derived from NatMK. Elongated pores, the result of unreacted metakaolin sheets becoming dislodged from the surface of the geopolymer, are clearly visible.

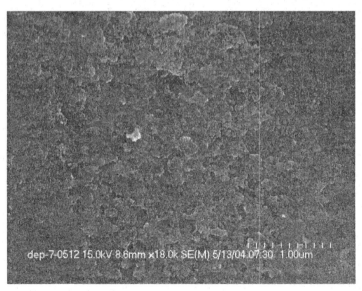

Fig.7: SEM micrograph of the surface of the synthetically derived geopolymer, showing homogeneous reactivity.

Bright field TEM of the synthetic geopolymer, Fig. 8, shows an amorphous structure of nanoparticulates (dark spotted areas) approximately 4 nm in diameter separated by mesoporosity or a secondary phase (lighter regions).

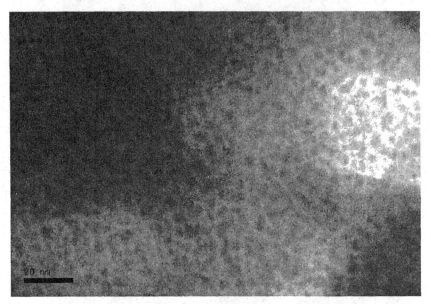

Fig. 8: TEM micrograph of synthetically derived geopolymer showing an amorphous, speckled microstructure, which is characteristic of natural geopolymers (bar = 20 nm).

DISCUSSION

SynMK is an excellent analog of natural metakaolin, but one that lacks the impurities intrinsically found in natural sources. Particle morphology and XRD patterns of the NatMK and SynMK matched well. XRD of the resulting synthetic geopolymer showed the same characteristic "halo" shift found in naturally derived geopolymers.[16]

The SynMK is highly reactive, requiring the use of an ice-water bath when the SynMK was mixed with alkali-silicate solutions. Although no quantitative measurements were made, the high reaction rate may be due to the high surface area of the SynMK, which would be contrary to the result that Rahier et al. reported using modulated differential scanning calorimetry (MDSC).[23] In their work, Rahier et al. reported that increasing the surface area of metakaolin particles above 14 m^2/g had no effect upon the reaction rate between metakaolin and alkali silicate solutions. Additional work is required to determine if the SynMK used in this study is intrinsically more reactive than NatMK, or if the high reactivity of SynMK is due to the high surface area of the SynMK.

In contrast to the similarities found in between NatMK and SynMK, the microstructure of the synthetically derived geopolymer appeared to be denser than that of the NatMK-based geopolymer. The higher density microstructure of the synthetic geopolymer may be due to the high surface area of the starting SynMK, which allows for more complete dissolution of the aluminosilicate particles. This result is similar to what Weng et al. reported.[25] As milling increased the surface area of the metakaolin in Weng et al.'s study, the microstructure of the final geopolymer became denser and had fewer unreacted metakaolin sheets present. Complete dissolution, which may have occurred in the case of the synthetically derived geopolymer would preclude unreacted metakaolin sheets in the final geopolymer microstructure.

Bright field TEM of synthetic geopolymers demonstrated that the morphology of the synthetic geopolymer is similar to that of a geopolymer derived from NatMK found in previous work.[6] The microstructure appeared to contain two distinct phases, dark regions ~4 nm in diameter, separated by a lighter continuous (or nearly continuous) phase. It is unclear if lighter regions found in bright field TEM images correspond to a secondary phase or mesoporosity.

CONCLUSIONS

SynMK with a surface area of 166 m^2/g and a median particle size of 1.62 μm was made using the organic steric entrapment method. SEM of SynMK showed sheet-like morphologies similar to NatMK. SynMK and NatMK were both reacted with potassium silicate solutions to create synthetically derived geopolymers. XRD indicated that SynMK and the corresponding synthetic geopolymer had the same characteristic diffraction patterns as NatMK and geopolymer based on NatMK. SEM analysis of the naturally and synthetically derived geopolymers showed that the synthetic geopolymer had a denser and more homogeneous microstructure than the microstructure of the naturally derived geopolymer. It is speculated that the high surface area of the SynMK particles allows them to dissolve more rapidly and completely when exposed to alkali-silicate solutions, resulting in a denser and more uniform microstructure. SEM showed no aluminosilicate sheets from the SynMK remaining in the synthetically derived geopolymer. Bright field TEM of the geopolymer based on SynMK showed a two-phase microstructure of dark regions ~4 nm in diameter separated by a lighter phase. The lighter phase may be solid material or mesoporosity. Compared to TEM results in literature of naturally based geopolymers, the synthetic geopolymer had a similar microstructure.

ACKNOWLEDGEMENTS

This work was supported by the AFOSR, under STTR Grant number F49620-02 C-010. This work was carried out using some of the facilities in the Center for Microanalysis of Materials, University of Illinois at Urbana-Champaign, which is partially supported by the U.S. Department of Energy under grant DEFG02-91-ER45439.

REFERENCES

[1] R.E. Lyon, P.N Balaguru, A. Foden, U. Sorathia, J. Davidovits and M. Davidovits, "Fire-Resistant Aluminosilicate Composites," *Fire Materials*, **21** [2] 67-73 (1997).

[2] J. Davidovits, M. Davidovits and N. Davodivits, "Geopolymer, Ultra High-Temperature Tooling Material for the Manufacture of Advanced Composites," *International SAMPE Symposium and Exhibition*, **36** [2] 1939-1949 (1991).

[3] C.G. Papakonstantinou, P.N. Balaguru and R.E. Lyon, "Comparative Study of High Temperature Composites," *Composites: Part B*, **32** 637-649 (2001).

[4] S.C. Förster, T.J. Graule and L.J. Gauckler, "Strength and Toughness of Reinforced Chemically Bonded Ceramics," in *Cement Technology*, edited by E.M. Gartner and H. Uchikawa. The American Ceramic Society. Ceramic Transactions, **40** 247-256 (1994).

[5] J. Davidovits, M. Davidovits and N. Davidovits, "Geopolymer, Ultra High-Temperature Tooling Material for the Manufacture of Advanced Composites," *International SAMPE Symposium and Exhibition*, **36** [2] 1939-1949 (1991).

[6] W. M. Kriven, J. L. Bell and M. Gordon, "Microstructure and Microchemistry of Fully-Reacted Geopolymers and Geopolymer Matrix Composites," in *Advances in Ceramic Matrix Composites IX*, Edited by N. P. Bansal, J. P. Singh, W. M. Kriven, H. Schneider. The American Ceramic Society. Ceramic Transactions, **153** 227-250 (1994).

[7] J.G.S. Van Jaarsveld, J.S.J. Van Deventer, L. Lorenzen, "The Potential Use of Geopolymeric Materials to Immobilize Toxic Metals: Part I. Theory and Applications," *Minerals Engineering* **10** [7] 659-669 (1997).

[8] J.G.S. Van Jaarsveld, J.S.J. Van Deventer, A. Schwartzman, "The Potential use of Geopolymeric Materials to Immobilise Toxic Metals: Part II. Material and Leaching Characteristics," *Minerals Engineering*, **12** [1] 75-91 (1999).

[9] J.W. Phair, J.S.J. Van Deventer, J.D. Smith, "Effect of Al Source and Alkali Activation on Pb and Cu Immobilisation in Fly-Ash Based "Geopolymers"," *Applied Geochemistry*, **19** [3] 423-434 (2004).

[10] H. Xu, Van Deventer J.S.J., "Geopolymerisation of Multiple Minerals," *Minerals Engineering*, **15** [12] 1131-1139 (2002).

[11] T.W. Cheng, J.P. Chiu, "Fire-Resistant Geopolymer Produced by Granulated Blast Furnace Slag," *Minerals Engineering*, **16** [3] 205-210 (2003).

[12] H Xu, J.S.J. Van Deventer, "Effect of Source Materials on Geopolymerization," *Industrial & Engineering Chemistry Research*, **42** [8] 1698-1706 (2003).

[13] J.G.S. Van Jaarsveld, J.S.J. Van Deventer, "Effect of the Alkali Metal Activator on the Properties of Fly Ash-Based Geopolymers," *Industrial & Engineering Chemistry Research*, **38** [10] 3932-3941 (1999).

[14] H. Xu, J.S.J Van Deventer, G.C. Lukey, "Effect of Alkali Metals on the Preferential Geopolymerization of Stilbite/Kaolinite Mixtures," *Industrial & Engineering Chemistry Research*, **40** [17] 3749-3756 (2001).

[15] H. Xu, J.S.J.Van Deventer, "The Geopolymerisation of Alumino-Silicate Minerals," *International Journal of Mineral Processing*, **59** [3] 247-266 (2000).

[16] V.F.F. Barbosa, K.J.D. MacKenzie, C. Thaumaturgo, "Synthesis and Characterisation of Materials Based on Inorganic Polymers of Alumina and Silica: Sodium Polysialate Polymers," *International Journal of Inorganic Materials* **2** [4] 309-317 (2000).

[17] V.F.F. Barbosa, K.J.D. MacKenzie, "Synthesis and Thermal Behaviour of Potassium Sialate Geopolymers." Materials Letters, **57** [9-10] 1477-1482 (2003).

[18] V.F.F. Barbosa, K.J.D. MacKenzie, "Thermal Behaviour of Inorganic Geopolymers and Composites Derived from Sodium Polysialate." *Materials Research Bulletin*, **38** [2] 319-331 (2003).

[19] J. Davidovits, "Geopolymers - Inorganic Polymeric New Materials," *Journal of Thermal Analysis*, **37** [8] 1633-1656 (1991).

[20] H. Rahier, B. Van Mele, M. Biesemans, J. Wastiels and X. Wu, "Low-Temperature Synthesized Aluminosilicate Glasses. Part I. Low-Temperature Reaction Stoichiometry and Structure of a Model Compound," *Journal of Materials Science*, **31** [1] 71-79 (1996).

[21] H. Rahier, B. Van Mele, J. Wastiels, "Low-Temperature Synthesized Aluminosilicate Glasses. Part II. Rheological Transformations during Low Temperature Glasses," *Journal of Materials Science*, **31** [1] 80-85 (1996).

[22] Rahier H, Simons W, VanMele B, Biesemans M. "Low-Temperature Synthesized Aluminosilicate Glasses. Part III. Influence of the Composition of the Silicate Solution on Production, Structure and Properties," *Journal of Materials Science*, **32** [9] 2237-2247 (1997).

[23] H. Rahier H, J.F. Denayer, B. Van Mele, "Low-Temperature Synthesized Aluminosilicate Glasses. Part IV. Modulated DSC Study on the Effect of Particle Size of Metakaolinite on the Production of Inorganic Polymer Glasses," *Journal of Materials Science*, **38** [14] 3131-3136 (2003).

[24] H. Xu, J.S.J Van Deventer, "Microstructural Characterisation of Geopolymers Synthesized from Kaolinite/Stilbite Mixtures using XRD, MAS-NMR, SEM/EDX, TEM/EDX, and HREM," *Cement And Concrete Research*, **32** [11] 1705-1716 (2002).

[25] L. Weng, K. Sagoe-Crentsil and T. Brown, "Speciation and Hydrolysis Kinetics of Aluminates in Inorganic Polymer Systems," *Geopolymers 2002*, Melbourne, Australia, Oct.28-29 (2002).

[26] J. P. Hos, P. G. McCormick, L. T. Byrne, "Investigation of a Synthetic Aluminosilicate Inorganic Polymer," *Journal of Materials Science*, **37** [11] 2311-2316 (2002).

[27] M. P. Pechini, "Method of Preparing Lead and Alkaline Earth Titanates and Niobates and Coating Method Using the Same to Form a Capacitor," U.S. Pat. No. 3 330 697, 1967.

[28] M. A. Gulgun and W. M. Kriven, "A Simple Solution-Polymerization Route for Oxide Powder Synthesis," *Ceramic Transactions* **62,** 57 – 66 (1995).

[29] S. J. Lee and W. M. Kriven, "Crystallization and Densification of Nano-size, Amorphous Cordierite Powder Prepared by a Solution-Polymerization Route," *Journal of the American Ceramic Society,* **81** [10] 2605-2612 (1998).

[30] M. A. Gulgun, M. H. Nguyen and W. M. Kriven," Polymerized Organic-Inorganic Synthesis of Mixed Oxides," *Journal of the American Ceramic Society,* **82** [3] 556-560 (1999).

[31] W. M. Kriven, "Synthesis of Oxide Powders via a Polymeric Steric Entrapment Precursor Route," *Journal of Materials Research*, **14** [8] 3417-3426 (1999).

[32] S. J. Lee and W. M. Kriven, "Preparation of Ceramic Powders by a Solution-Polymerization Route Employing PVA Solution," *Ceramic Engineering and Science Proceedings,* **19** [4] 469-476 (1998).

[33] E. A. Benson, S. J. Lee and W. M. Kriven, "Preparation of Portland Cement Components by PVA Solution Polymerization," *Journal of the American Ceramic Society,* **82** [8] 2049-2055 (1999).

[34] S. J. Lee, M D. Biegalski and W. M. Kriven, "Barium Titanate and Barium Orthotitanate Powders through an Ethylene Glycol Polymerization Route," *Ceramic Engineering and Science Proceedings,* **20** [3] 11-18 (1999).

[35] ," S. J. Lee and W. M. Kriven," A Submicron-Scale Duplex Zirconia and Alumina Composite by Polymer Complexation Processing," *Ceramic Engineering and Science Proceedings,* **20** [3] 69-76 (1999).

[36] S. J. Lee, M D. Biegalski and W. M. Kriven , "Powder Synthesis of Barium Titanate and Barium Orthotitnate via an Ethylene Glycol Polymerization Route," *Journal of Materials Research,* **14** [7] 3001-3006 (1999).

[37] W. M. Kriven, S. J. Lee, M. A. Gulgun, M. H. Nguyen and D. K. Kim, "Synthesis of Oxide Powders via Polymeric Steric Entrapment," in *Innovative Processing/Synthesis: Ceramics, Glasses, Composites III*, edited by J. P. Singh, N. P. Bansal, K. Niihara. The American Ceramic Society. Ceramic Transactions, **108**, 99-110 (2000).

CHEMICALLY BONDED PHOSPHATE CERAMICS – A NOVEL CLASS OF GEOPOLYMERS

Arun S. Wagh
Argonne National Laboratory
9700 S. Cass Avenue
Argonne, IL 60439

ABSTRACT

Geopolymers are cements or ceramics that mimic natural minerals. They are formed by polymerization of inorganic molecules containing aluminum, silicon, oxygen, and other elements. Under this broad definition, chemically bonded phosphate ceramics may also be considered as a new class of geopolymers. Phosphate ceramics are synthesized at room temperature, and they set rapidly like conventional polymers. They contain naturally occurring mineral phases, notably apatite. They are formed by an acid-base reaction between a metal oxide and an acid phosphate. Virtually any divalent or trivalent oxide that is sparingly soluble may be used to form these ceramics. They are hard and dense, and exhibit superior mechanical properties compared to conventional cement. They have found a wide range of applications such as dental cements, construction materials, oil well cements, and hazardous and radioactive waste stabilization. This paper summarizes the kinetics of formation of these phosphate geopolymers, their phase composition and unique properties, microstructure, and potential niche applications. Emphasis will be on apatite-forming geopolymers and their key role in various applications.

INTRODUCTION

Geopolymers are aluminosilicate cements or ceramics that are formed at room temperature. They are produced by pyroprocessing of naturally occurring kaolin (alumina-rich clay) into metakaolin [1]. This metakaolin is then reacted with an alkali hydroxide or sodium silicate to yield a rock-like hard mass. Geopolymers are also formed by reacting sodium-silicate-loaded fly ash with water [2], and also from aluminosilicate minerals. Thus, the blend that forms geopolymers contains alkaline material, and, in that respect, is similar to conventional Portland cement, but geopolymers exhibit properties superior to conventional cements.

The word "geopolymers" was originally defined by Davidovits [3] as materials formed by tetrahedral aluminate and silicate units condensed at ambient temperature and

* Work is supported by U.S. Department of Energy, Office of Technology Development, under Contract No. W-31-109-Eng-38.

charge balanced by the presence of monovalent alkali metal ions. This definition is based on the structure of the materials. The word geopolymer is also loosely used to mean materials formed at room temperature, that exhibit ceramic-like properties and are x-ray amorphous [4]. Thus, the definition can be based on the structure or can be more utilitarian. The word "Geo" implies these minerals mimic natural minerals, possibly clay. Because they are made of long-chain molecules of silicates, they may be considered inorganic polymers. Thus, together the name becomes "geopolymers."

This more practical definition may be extended to other naturally occurring inorganic polymers, such as phosphates, vanadates, borates, and arsenates. In particular, phosphates can be synthesized at a laboratory or industrial scale and are considered chemically bonded ceramics. Just as silica tetrahedron is the basic unit of conventional geopolymers, chemically bonded phosphate ceramics (CBPCs) have $(PO_4)^{3-}$ tetrahedron as the basic building block. They are formed at room temperature. Though their structures are not well studied, they invariably contain amorphous phases that seem to form the binding phases, especially when one uses a source of amorphous silica in them such as fly ash, and they also exhibit properties superior to cement [5]. These materials are now finding applications in radioactive and hazardous waste stabilization, architectural products, oil well cements, and denstry. The motivation behind the current interest in conventional geopolymers is also waste management and structural materials. Unlike conventional geopolymers, however, the structure of CBPCs is better understood. The purpose of this overview is to bring to the attention of researchers the importance of the CBPCs as novel geopolymers and initiate an interest in further research.

Syntheses of Phosphate Geopolymers

The main difference between the silicate based geopolymers and phosphate geopolymers, however, is their syntheses. Silicate based geopolymers are synthesized in alkaline environment, but phosphate geopolymers are fabricated by acid-base reactions. Compared to conventional geopolymers, a very wide range of CBPCs may be synthesized by acid-base reaction between an inorganic oxide (preferably that of divalent and trivalent metals) and an acid phosphate. The reaction product is generally a hydrophosphate or an anhydrous phosphate that consolidates into a ceramic. The following are the most common examples [5-7]:

$$2CaO + Ca(H_2PO_4)_2 + H_2O => CaO + 2CaHPO_4 \cdot H_2O => Ca_3(PO_4)_2 + 2H_2O \tag{1}$$

$$MgO + KH_2PO_4 + 5 H_2O = MgKPO_4 \cdot 6H_2O \ (Ceramicrete^{TM}). \tag{2}$$

These reactions occur at room temperature. By controlling the rate of reaction, ceramics can be formed. With trivalent oxides, similar ceramics can be formed at a slightly elevated temperature. A good example is berlinite ($AlPO_4$), which is formed by the reaction between alumina and phosphoric acid:

$$Al_2O_3 + 2H_3PO_4 = 2AlPO_4 + 3H_2O. \tag{3}$$

Wagh et al. [8] have also demonstrated that CBPCs of trivalent oxides such as Fe_2O_3 and Mn_2O_3 may be produced by reduction of the oxide and then acid-base reaction of the reduced oxide with phosphoric acid. The reaction may be described by the following equation:

$$X_2O_3 + X + 3H_3PO_4 + nH_2O = 3XHPO_4 \cdot (n+3)H_2O. \qquad (4)$$

X in this equation is either Fe or Mn. This metal is used as the reductant, which partially reduces X_2O_3 and makes it soluble in the acid solution, after which the acid-base reaction is initiated. With these three routes of producing ceramics, any divalent or trivalent oxide may be used to form a ceramic at ambient or low temperature, in the same manner as a conventional geopolymer is made. However, acid-base reactions form CBPCs, and the end product is nearly neutral, while conventional geopolymers are formed in highly alkaline solutions.

These simple reactions are the basis for producing more complex and longer chain molecules and polymeric structures (Table 1). The most important phosphate minerals formed this way are apatites, which have been synthesized in a laboratory and have been formed in phosphate reactions for waste management operations, dental cements, and biomaterials. Some of these minerals have also become components of superior structural materials proposed for architectural and construction applications. Because apatites are commonly used as CBPCs, we will discuss them in more detail.

Apatite structures are given by $M_{10}(PO_4)_6Z_2$, where M stands for metal such as Ca, Mg, or Pb, and Z is OH, Cl, F, or CO_3. Correspondingly, the apatites are named as hydroxyapatite for Z=OH, chloroapatite for Z=Cl, fluoroapatite for Z=F, and carbonate apatite for Z=CO_3. Calcium based apatites are the most common. They are found in natural minerals, bones and modern-day bioceramics, and now are becoming an integral part of minerals formed in waste management operations and other man-made ceramics.

Table 1 lists some of the specific CBPC minerals.

Table 1. Phosphate minerals found in various ceramic applications [6-9]

Mineral	Formula	Application
Hydroxyapatite	$Ca_5(PO_4)_3 \cdot OH$	Bioceramics
Chloroapatite	$Ca_5(PO_4)_3 \cdot Cl$	Waste management
Fluoroapatite	$Ca_5(PO_4)_3 \cdot F$	Bioceramics
Carbonate apatite Dahllite	$Ca_5(PO_4)_3 \cdot CO_3$ $Ca_{8.8}(HPO_4)_{0.5}(PO_4)_{4.5}(CO_3)_{0.7}(OH)_{1.3}$	Bioceramic
Hydroxypyromorphite	$Pb_5(PO_4)_3 \cdot OH$	Waste management
Chloropyromorphite	$Pb_5(PO_4)_3 \cdot Cl$	Waste management
Fluoropyromorphite	$Pb_5(PO_4)_3 \cdot F$	Waste management
Mg-based phosphates Ceramicrete Struvite Mg-Cs phosphate	$MgXPO_4 \cdot nH_2O$, X = K, NH_4 $MgKPO_4 \cdot nH_2O$ $Mg(NH_4)PO_4 \cdot 22H_2O$ $MgCsPO_4 \cdot 6H_2O$	Structural material, Rad waste and Fission product encapsulation

Kinetics of Formation of Phosphate Geopolymers

The step-by-step kinetics of formation of CBPCs is illustrated in Fig. 1 and summarized below.

1. Dissolution of oxides and formation of sols by hydrolyses

Sparsely soluble metal oxides, when stirred into an acid solution, dissolve slowly and release cations and oxygen-containing anions (Fig. 1a, dissolution step). The cations react with water molecules and form positively charged "aquasols" by hydrolysis (Fig. 1b, hydration step). The dissolution and hydrolysis are the controlling steps in forming CBPCs.

2. Acid-base reaction and gel formation

The sols subsequently react with the aqueous phosphate anions (Fig. 1c) to form the hydrophosphate salts, while the protons and oxygen react to form water. The newly formed hydrophosphate salts form a network of molecules in the aqueous solution that leads to gel formation (Fig. 1d).

3. Saturation and crystallization of the gel into a ceramic

As the reaction proceeds, this process introduces more and more reaction products into the gel, and it thickens. When sufficiently thickened, the gel crystallizes around the unreacted core of each grain of the metal oxide into a well-connected crystal lattice that grows into a monolithic ceramic (Fig. 1c).

a. Dissolution of oxide

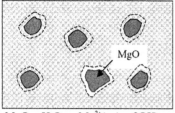

$$MgO + H_2O \rightarrow Mg^{2+}(aq) + 2OH^-$$

b. Formation of aquosols

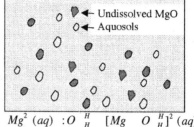

$$Mg^2 \ (aq) \ :O \ {}^H_H \ \ [Mg \ \ O \ {}^H_H]^2 \ (aq)$$

c. Acid-base reaction and condensation

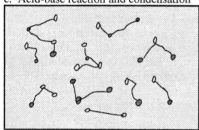

$$Mg(OH)^{2+}(aq) + HPO_4^{2-} + 2H_2O$$
$$\rightarrow MgHPO_4 \cdot 3H_2O$$

d. Percolation and gel formation

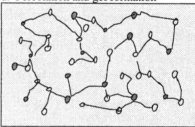

e. Saturation and crystallization

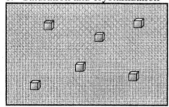

Fig. 1. Pictorial representation of formation of chemically bonded phosphate ceramic.

APPLICATIONS

As listed in Table 1, the phosphate geopolymers are already in use in several applications. Dental cements are the first geopolymers to find a market. Some of these cements were developed as early as in the late 19th and beginning of the 20th century [9, 13]. They are still being perfected to produce hydroxy- and chloro-apatites so they are biocompatible and can be used as bioceramics. The most recent bioceramics developed by Constantz et al [8], mimic bone. The product contains a form of natural mineral dahllite and has the formula,

$$Ca_{8.8}(HPO_4)_{0.5}(PO_4)_{4.5}(CO_3)_{0.7}(OH)_{1.3}.$$

Such complicated minerals and their ceramic that mimics natural bone have been produced by simple acid-base reactions at room temperature. Clinical tests have been completed on this formulation and it is a commercial product now [8].

The other major application is also a commercial process: stabilization of hazardous waste by converting contaminants, such as Pb, Cd, Zn, and Cr, into their respective apatite minerals. Apatite minerals are ideal forms for stabilizing hazardous contaminants, because they are extremely insoluble and do not leach into the groundwater. In addition, because most hazardous waste streams contain anions such as Cl, and F, they can also be stabilized in corresponding apatite forms, such as chloro- and fluoroapatites. Detailed study by several researchers in the last ten years has shown that hazardous contaminants such as chloropyromorphite stabilize conaminants such as $PbCl_2$.

Eighmy and Eusden [7] surveyed recent literature on phosphate amendment of various industrial waste streams. They found 39 patents in this area since 1994. An updated summary of the various waste streams, either treated or tested by phosphate amendment or solidification, is given in Table 2. The key reactions in these processes are given by

$$Ca_5(PO_4)_3(OH) + 7H+ = 5Ca^{2+} + 3 H_2PO_4^- + H_2O, pKsp = 15.8 +/-1.7 \qquad (16.1)$$

$$PbCO_3 + H^+ = Pb^{2+} + HCO_3^- \qquad (16.2)$$

$$5 Pb^{2+} + 3 H_2PO_4^- + Cl^- = Pb_5(PO_4)_3Cl, \ pKsp = 18.9. \qquad (16.3)$$

These acid-base reactions raise the pH of the waste stream in which chloropyromorphite is formed. Because this compound is extremely insoluble [pKsp = 84.4], it precipitates out.

Eighmy and Eusden [7] have extended the concept of stabilizing hazardous metals to Cd and Zn into their corresponding apatite minerals such as $Cd_5(PO_4)_3Cl$, $Cd_5(PO_4)_3OH$, $Zn_5(PO_4)_3Cl$, and $Zn_5(PO_4)_3OH$. In addition to the apatite forms such as the pyromorphites, the authors have suggested stable forms of other phosphates, such as the tertiary metal phosphates $Pb_3(PO_4)_2$, $Cd_3(PO_4)_2$, and $Zn_3(PO_4)_2$, and even the tetra metal phosphate of Pb, $Pb_4O(PO_4)_2$. Furthermore, these authors have also identified much more complex solid-solution phases of these minerals with phosphates of Fe, Ca, and Al.

Table 2. Phosphate-treated or -tested hazardous waste streams, target contaminants, and number of references published in last ten years

Waste stream	References
Municipal solid waste and other ash	6
Contaminated soils	22
Contaminated sediments	2
Electric arc furnace dust	1
Smelter slags	2
Mine tailings	1
Heavy-metal-plating sludge	1
Industrial waste water	2
Heavy metal contaminated sludge capping	2
U-contaminated groundwater and soil	4

The concept of stabilizing hazardous waste by using phosphates has been generalized by Wagh et al [8] to stabilize radioactive contaminants. Table 3 lists the various radioactive contaminants in waste streams and summarizes how they are stabilized in a phosphate matrix.

In all cases, magnesium phosphate is used to produce a matrix in which the contaminants are stabilized as substitutional impurities, and the anions are stabilized within appropriate apatites. There is evidence that most soluble compounds, viz., sodium nitrate, can also be stabilized in such a system. We are in search of a nitroapatite that stabilizes this compound.

The phosphate geopolymers are also applicable as oil field cements. Sugama and his group [9], as well as Wagh and his group [10] have developed a range of formulations to produce phosphate-based cements. Sugama and his group concentrated on geothermal wells, where the temperature is > 250°F, but setting, curing, and overall durability of the cement may be affected by carbonation of downhole gases. Wagh and his group, on the other hand, produced phosphate geopolymer formulations for the entire range of wells, from permafrost to deep wells, where the temperature is > 300°F. Boric acid has been used to delay the rate of reactions and produce cements that can be pumped at the desired depths. Table 4 provides the different formulations developed for oil-field applications.

Table 3. Radioctive contaminants and their solubility

Contaminants	Form in the waste	Solubility
Actinides	Generally present in fully oxidized form, but salts such as nitrates and chlorides are present in solution.	Fully oxidized forms are insoluble. Oxides of lower oxidation states may be slightly soluble and can be stabilized as monazites. Actinide salts are soluble and may be stabilized as apatites.
Radioactive fission products	Cesium (Cs), strontium (Sr), barium (Ba), and technetium (Tc) are main isotopes present as salts (nitrates and chlorides). Tc may be present as oxides.	All salts are soluble. Tc oxidizes as soluble pertechnetate, but in reduced state, it is less soluble. Cs is stabilized in as $MgCsPO_4 \cdot nH_2O$ and Sr and Ba as orthophosphates. Stabilization of Tc is very effective but exact mineral form is not known.
Radium	Ra disintegrates into Rn. Rn is a gas and hence is of main concern during stabilization.	Ra is soluble. Possibly Ra is stabilized as orthophosphate and then encapsulated in the phosphate matrix.
Salt components	Sodium and potassium nitrate, chloride, and sulfates.	Cations are stabilized in orthophosphates, and anions in corresponding apatite forms.

Table 4. CBPC oil-well slurry compositions tested at ANL and BNL [7, 9-10]

Well Conditions	Pumping T (°F)	Components (in g)
Permafrost and ambient	32 and 70	MgO reacted with KH_2PO_4. $CaSiO_3$ and Class C fly ash are the extenders.
Shallow	120 - 200	MgO reacted with KH_2PO_4. Class C fly ash is the extender.
Deep well and geothermal	250 - 350	Calcined alumina reacted with H_3PO_4 solution. Aluminum hydroxide is the activator.
Geothermal	> 250	Calcium aluminate cement reacted with sodium metaphosphate, and fly ash is the extender.

The cements that work for all these niche applications are also suitable for structural materials. For example, the Mg-based cements have been used as road repair materials, architectural facades, and bridge surfacing. Properties of these cements are reported in Ref. [12].

CONCLUSIONS

This overview suggests chemically bonded phosphate ceramics may also be considered a class of geopolymers in a wider definition of the word "geopolymers." The structure of some of the chemically bonded phosphate ceramics may also be similar to that of silicate geopolymers. This similarity is because silicate geopolymers are formed from the building block of silicate tetrahedron, while phosphate geopolymers are based on phosphate tetrahedron. In fact, silica and aluminum phosphate structures are isomorphous. Though our generalization of the definition of geopolymers is based on the formation process and properties, it is likely that the two products may also be similar in structure, and this possibility needs to be investigated. Compared to conventional silicate geopolymers, chemically bonded phosphate ceramics are better understood, have found commercial applications, and yet, have failed to be recognized by the geopolymer scientific community. Further research in geopolymers should take these promising materials into consideration.

ACKNOWLEDGMENTS

The work was supported by U.S. Department of Energy, Office of Technology Development, under Contract No. W-31-109-Eng-38.

REFERENCES
1. Glasser, F. P., Cements from Micro to Macrostructures, Ceram. Trans. J., 89 [6] (1990) 195-202.
2. Grutzeck, M., Zeolite Formation in Alkali-Activated Cementitious Systems, Cement and Concrete Research, 34 [6] (2004) 949-955.
2. Davidovits, J., Geopolymers: Inorganic Polymeric New Materials, J. thermal Analysis 37[8](1991) 1633-1656.
3. MacKenzie, K. J. D., What Are These Things Called Geopolymers? A Physicochemical Perspective, Advances in Ceramic Matrix Composites IX, N. P. Bansal, J. P. Singh, W. M. Kriven, and H. Schneider (eds.) 153 (2003) 175-186.
4. Wagh, A. S., and S. Y. Jeong, Chemically Bonded Phosphate Ceramics: I. A Dissolution Model of Formation, J. Ceram. Soc., 86 [11] (2003) 1838-1844.
5. Wagh, A. S., and S. Y. Jeong, Chemically Bonded Phosphate Ceramics: II. Warm-Temperature Process for Alumina Ceramics, J. Ceram. Soc., 86 [11] (2003) 1845-1849.
6. Wagh, A. S., and S. Y. Jeong, Chemically Bonded Phosphate Ceramics: III. Reduction Mechanism and Its Application to Iron Phosphate Ceramics, J. Ceram. Soc., 86 [11] (2003) 1850-1855.
7. Chow, L. C., Calcium Phosphate Cements: Chemistry, Properties, and Applications, in Mat. Res. Soc. Proc. 599 (2000) 27 – 37.

8. Eighmy, T. T., and J. Eusden, Phosphate Stabilization of MSW Combustion Residues: Geochemical Principles, to appear in *Energy, Waste, and Environment – A Geochemical Perspective*, Geological Society of America (to be published).
9. Constantz, B. R., et al., Skeletal Repair by in Situ Formation of the Mineral Phase of Bone, Science, 67 (1995) 1796 - 1799.
10. Wagh, A. S., and S. Y. Jeong, Chemically Bonded Phosphate Ceramics for Stabilization and Solidification of Mixed Waste, *Handbook of Mixed Waste Management Technology*, CRC Press, Boca Raton, FL, Chapter 6.3 (2000).
11. Sugama, T., Hot Alkali Carbonation of Sodium Metaphosphate Modified Fly Ash/Calcium Aluminate Blend Hydrothermal Cements, Cem. Concr. Res. 26 [11] (1996) 1661-1672.
12. Wagh, A. S., Chemically Bonded Phosphate Ceramic Borehole Sealants, Final report to Global Petroleum Research Institute, Argonne National Laboratory (2002).
13. American Association of State Highway and Transportation Officials, Laboratory Evaluations of Rapid Set Concrete Patching Materials, Report 99 NTPEP 160 (1999).
14. Wilson, A. D., and J. W. Nicholson, *Acid-Base Cements*, Cambridge University Press, Cambridge, U.K. (1993).

Mechanical Properties
and Micromechanical Modeling

FRACTURE AND CRACK GROWTH IN CERAMIC COMPOSITES AT HIGH TEMPERATURES

Raj N. Singh
Department of Chemical and Materials Engineering
University of Cincinnati, P.O. Box 210012
Cincinnati, OH 45221-0012

ABSTRACT

Fracture and crack growth resistance are important properties for applications of ceramic .composites. Consequently, the fracture and crack growth behaviors in two types of fiber-reinforced ceramic matrix composites (CMCs) are studied over a range of temperatures from 25°-1400°C. The crack is initiated and propagated from a notch and crack length is obtained in situ using a video system in order to study the crack propagation and fracture resistance accurately. The results from both composite systems indicated a rising R-curve behavior over the temperatures of this study. The experimental results are modeled using an analytical approach, which showed good agreement with the measured data. The analytical approach also proved to be very useful in not only predicting fracture resistance of CMCs but also for designing composites with superior properties for applications at high temperatures.

INTRODUCTION

Ceramic matrix composites (CMCs) are being considered for high temperature structural applications due to their high toughness and strength when compared with that of the monolithic ceramics. This combined with other attractive properties of ceramics such as low density, low thermal conductivity, as well as high wear and creep resistance, make them attractive candidates for use at elevated temperatures. In ceramic composites, the toughened behavior is caused primarily by the bridging fibers, which remain intact after matrix cracking [1,2]. The bridging fibers impose a closure stress on the crack wake, which opposes the far field applied stress thereby enhancing toughness. The fiber bridging stress is controlled by the mechanical properties of the fiber and matrix, and interfacial properties, especially the interfacial shear stress [3-6].

One common method for evaluating the toughening potential of ceramic materials is to measure the fracture resistance curve (R-curve), which relates the toughness to the crack length. However, measurement of crack length *in situ* at elevated temperatures is difficult. Consequently, most studies relied on unloading the sample to get at the crack length. This approach is not accurate because crack can close upon unloading and heal at high temperatures. In this study, the crack growth and fracture resistance behaviors of a unidirectional silicon carbide (SCS-6) fiber-reinforced zircon ($ZrSiO_4$) and a woven SiC_f-SiC composites were determined at elevated temperatures by using a video recording system to measure the relation between *in situ* crack length and applied load. These results were then used to calculate and model the R-curve behavior of composites for predicting their fracture resistance and for designing superior composites.

EXPERIMENTAL PROCEDURE

Two types of composites were used in this study. In one case, a silicon carbide fiber (SCS-6) reinforced zircon matrix composite was used. This composite was fabricated by uniaxially aligning SiC fibers into a preform and then incorporating the zircon matrix by tape casting and lamination techniques. The green body was then consolidated by hot pressing at about 1640 C in a flowing nitrogen atmosphere. The final fiber volume fraction in this composite was approximately 25%. In the second case, a woven SiC_f (Nicalon^{TM})-SiC matrix composite was used. This composites was fabricated by a melt-infiltration (MI) process in which Si melt was used to infiltrate a woven fiber perform. The total fiber content in this woven composite was about 40% (~20% normal to crack propagation). Further details on processing of these composites can be found in the literature [7-10].

The three point bending test was employed to propagate the crack from a notch and measure the fracture resistance as shown in Fig. 1. Specimens of dimensions 30 x 5 x 1.3 mm were tested with an outer span of 20 mm. A notch length of about 1.75 mm (an initial a/W of 0.35) was created using a fine diamond saw. Tests were performed using a servo hydraulic testing machine between room temperature and 1400 C in a flowing high purity argon atmosphere. Crosshead displacement rate was selected to be 0.1 μm/sec so that the crack length could be measured *in situ* by an optical method with a resolution of about 2.5 μm. This approach was used to measure the crack length so that an *in situ* relation between the applied load and the crack length could be obtained. This way of testing created an initiation and propagation of a single dominant crack from the notch at all temperatures, which minimized the errors in crack length measurements.

The experimental R-curve was determined from the applied stress intensity factor, K_{app}, using the ASTM standard equation [10]. For the sample with a span to specimen width ratio of 4, this equation is given as

$$K_{app} = \frac{3PS}{2BW^2} \sqrt{a} \, \frac{1.99 - \frac{a}{W}(1 - \frac{a}{W})[2.15 - 3.93\frac{a}{W} + 2.7(\frac{a}{W})^2]}{(1 + 2\frac{a}{W})(1 - \frac{a}{W})^{3/2}} \tag{1}$$

where P is the applied load, B is the specimen thickness, W is the specimen width, S is the span used in three point bending, and a is the crack length.

RESULTS AND DISCUSSION

Crack Propagation and Fracture Resistance of CMCs

Figure 2 shows the load-displacement responses of a SiC_f–Zircon composite measured between 25-1400°C. It shows the typical toughened ceramic matrix composite behavior, with a high load bearing capability above the matrix cracking stress. In a CMC, both the matrix and the fiber have low inherent fracture toughness due to the lack of crack tip plasticity, and the matrix

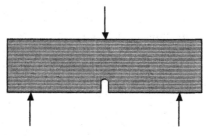

Fig. 1. Schematic showing the notch and load application in 3-point flexure mode used for crack propagation in CMCs.

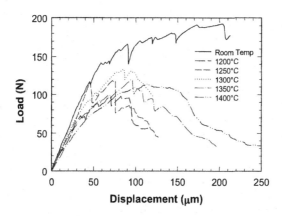

Fig. 2. Load-displacement data during crack propagation in SiC$_f$-Zircon composite between 25-1400°C.

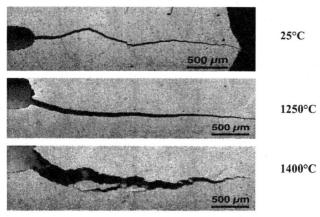

Fig. 3. The nature of crack propagation in SiC$_f$-Zircon composite at 25, 1250, and 1400°C.

has a low strain to failure. However, in a weakly bonded composite (where the interfacial shear strength between the matrix and the fiber is low), a matrix crack would be deflected at the fiber-matrix interface, leaving the fibers intact. The fibers can support the load even when the matrix has completely failed leading to toughened CMC response. The tests were stopped as soon as crack branching occurred or until the specimen failed depending on the mode of deformation. Load-displacement curves in Fig. 2 show that the fracture energy, a measure of toughness (area under the curve) decreases with increasing temperature.

Figure 3 shows the nature of crack propagation in CMCs between 25-1400°C. Fairly straight cracks initiated and propagated up to 1250°C with little evidence of matrix plasticity. The crack opening displacement increased with temperature up to 1250°C. Above 1250°C both fibers and zircon matrix in this CMC begin to show plasticity and the nature of crack propagation shown in Fig. 3 at 1400°C clearly displays a significant crack opening and matrix plasticity. In spite of this, the crack propagation even at 1400°C is primarily via a single dominant crack. Figure 4(a) shows the measured in situ load and crack length as a function of the displacement of the loading points for crack propagation at 25°C. The crack growth starts at loads much below the ultimate load carrying capability of the composite and continues with increasing loads until the ultimate load is reached. This type of load vs crack propagation behavior is observed up to 1250°C. However, at temperatures above 1300°C most of the crack propagation occurs at or above the ultimate in load carrying capability of the composites. An example of this behavior at 1400°C is shown in Fig. 4(b). This behavior is attributed to the onset of significant creep deformation in the matrix phase as well as lowered fiber strength above 1300°C [8]. Consequently, the crack growth behaviors are different above and below ~1250°C in this uniaxial SiC_f–Zircon composite system.

A summary of crack propagation results for SiC_f–Zircon CMC between 25-1400°C is given in Fig. 5 (a). The experimental results of fracture resistance calculated using Eq. (1) and in situ crack growth data from Fig. 5(a) are shown in Fig. 5 (b) for the temperature range of 20 to 1400°C. Figure 5 (b) shows an initial toughness of 6~10 MPa·m$^{1/2}$ and a continuous increase of fracture resistance value with crack propagation, a/W, at all the temperatures. The fracture resistance curves at 1200° and 1250°C are identical to each other. Although the initial toughness was approximately the same in the temperature range of 1200°-1250°C as that measured at room temperature, the slope of the crack resistance curves showed a marked decrease for $a/W > 0.5$ in comparison with the results at room temperature. This reduction in the slope of the fracture resistance as a function of a/W was especially significant for specimens tested at 1350° and 1400°C. These results show that the crack growth resistance of CMC decreases above 1250°C because of the lower strength of the reinforcing fibers. In spite of this, the CMC displayed increasing fracture resistance with crack growth or a rising K_R response at all temperatures of this study.

The SiC_f–SiC composite fabricated by melt infiltration (MI) also displayed crack growth via a dominant single crack as shown in Fig. 6 between 25-1315°C. However, in this case most of

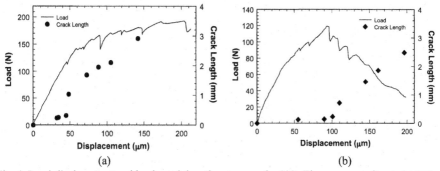

(a)   (b)

Fig. 4. Load-displacement and load-crack length responses for SiC$_f$-Zircon composite at (a) 25° and (b) 1400°C showing load required for crack growth.

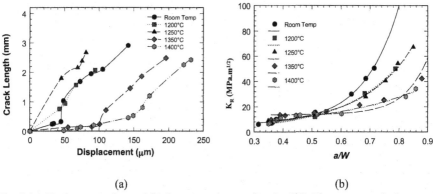

(a)   (b)

Fig. 5. (a) Crack propagation and (b) fracture resistance data for SiC$_f$-Zircon composite between 25-1400°C.

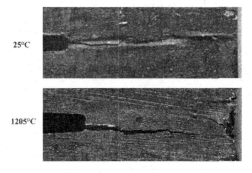

Fig. 6. The nature of crack propagation in SiC$_f$-SiC MI-composite at 25° and 1205°C.

Load Vs Crack Length (0 orientation)

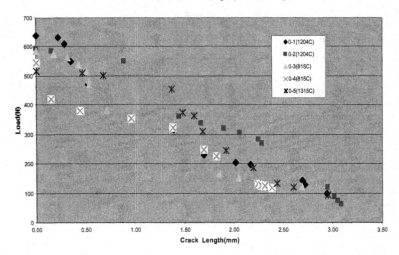

Fig. 7. The dependence of crack propagation on applied load in SiC$_f$-SiC MI-composite between 815° and 1315°C.

K$_R$ Vs a/w (0 Orientation)

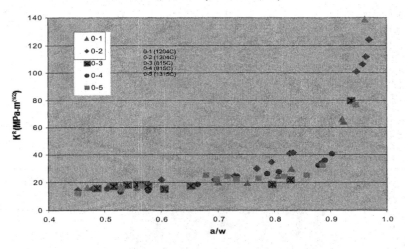

Fig. 8. The dependence of fracture toughness on crack propagation in SiC$_f$-SiC MI-composites between 815° and 1315°C.

the crack growth occurs at loads near or above the ultimate in load carrying capability of the composite because of the strong matrix of SiC. As a consequence, the crack growth occurs with decreasing load as shown in Fig. 7 for data between 815-1315°C. These in situ crack propagation data are then used to calculate the fracture resistance responses as shown in Fig. 8, which clearly show a slowly rising K_R behavior with crack growth up to a/W value of about 0.7 and then a steep rising K_R behavior with additional crack growth. This K_R behavior for woven MI CMC is very similar to the behavior for the uniaxial Zircon-SiC$_f$ uniaxial CMC presented earlier in Fig. 5(b) indicating a generic crack growth resistance response for CMC over a range of temperatures. It also suggests that the in situ approach of measuring crack growth and fracture resistance used in this study can be very useful in assessing fracture resistance of composites over a range of temperatures. Consequently, these results provided the encouragement to model the fracture resistance of CMCs using an analytical approach for providing predictive capability for CMCs.

Theoretical Fracture Resistance Curve

For the numerical calculation, a well-developed bridging stress law, which considers both debonding and bridging of intact fiber and fiber pullout processes, and a numerical approach developed by LLorca et al. [7,11-13] were used in this study to determine the theoretical R-curve behavior in ceramic composites. A more detailed approach can be found in the literature [7,12]. The basic equation describing the bridging contribution to the stress intensity factor is expressed by:

$$K_R = K_M + \int_{a_o}^{a} \frac{2\sigma_b(u)}{\sqrt{\pi a}} H(\frac{a}{W}, \frac{x}{a}) \, da \qquad (2)$$

where K_M is the matrix toughness, $\sigma_b(u)$ is the bridging stress function, u is the crack opening displacement, a_o is the initial crack length, and H is the weight function depending on the specimen geometry and loading configuration. For three point bending geometry used in this study, the weight function H is described in the appendix of a previous work [7].

There are several theoretical derivations to obtain the $\sigma_b(u)$ function: (i) a "hardening" function, $\sigma_b \propto u^{1/2}$, was derived by considering debonding between fibers and matrix with a constant interfacial shear stress and fiber strength [5,14,15], (ii) a "softening" function, which used a simple approach of a monotonically decreasing function and considered the fiber pull-out effect [2,16-18], and (iii) "hardening" and "softening" function, based on a more sophisticated analysis by Thouless and Evans [3], who considered both the fiber bridging and fiber pull-out processes, and further combined the statistical distribution of fiber strength.

The results by Thouless and Evans [3] showed that the $\sigma_b(u)$ function exhibited an initial increase with u due to the bridging fibers and then it decreased as fibers fractured and pulled out from the matrix. Their approach was used in this study by applying the following bridging stress-crack opening displacement function, $\sigma_b(u)$,

$$\frac{\sigma_b(u)}{V_f \Sigma} = U^{1/2} \exp(-U^{(m+1)/2}) + [1 - \exp(-U^{(m+1)/2})]$$

$$\times \left[\gamma \left\{ \frac{m+2}{m+1}, U^{(m+1)/2} \right\} - \frac{\Sigma(m+1)}{2E_f} U \right] \times \frac{1}{(1+\xi) \times (m+1)} \tag{3}$$

where $U = u/u_n$ is the normalized displacement, m is the Weibull modulus for fiber strength distribution, V_f and E_f are the volume fraction and elastic modulus of fibers, respectively, γ (α, β) is the incomplete gamma function, and the parameters ξ, Σ and u_n are defined by:

$$\xi = \frac{E_f V_f}{E_m V_m}, \tag{4a}$$

$$\Sigma = \frac{\sigma_f}{\Gamma(\frac{m+2}{m+1})}, \tag{4b}$$

$$u_n = \frac{\Sigma^2 R}{4\tau E_f (1+\xi)}, \tag{4c}$$

where V_m and E_m are the volume fraction and elastic modulus of the matrix, respectively, σ_f is the fiber strength and R is the radius of the fiber.

However, the bridging stress function of Eq.(3) does not take into account the presence of residual stresses, which are created by the thermal expansion mismatch between the fibers and matrix. The residual stress can change the bridging stress function and thus affect the toughness of the composite [5,19]. Considering the bridging stress function of Eq.(3), the first term on the right indicates the effect of the bridging process from intact fibers and the second term signifies the contribution of pullout process from broken fibers. The influence of the residual stress should only be considered for the first term since the residual stress is released upon fiber fracture. As a result, Eq.(3) can be modified and re-written as [7]:

$$\frac{\sigma_b(u)}{V_f \Sigma} = (U^{1/2} + \frac{\sigma_r^f E_c}{(1-V_f)E_m \Sigma}) \exp(-U^{(m+1)/2}) + [1 - \exp(-U^{(m+1)/2})]$$

$$\times \left[\gamma \left\{ \frac{m+2}{m+1}, U^{(m+1)/2} \right\} - \frac{\Sigma(m+1)}{2E_f} U \right] \times \frac{1}{(1+\xi) \times (m+1)} \tag{5}$$

where σ_r^f is the residual stress in the fiber along longitudinal direction.

An iterative numerical procedure was employed to calculate the theoretical R-curve [7,11-13] by using room and elevated temperature mechanical properties of SiC$_f$, Zircon, and SiC$_f$/Zircon composites already known from our extensive studies on this composite system. In addition, the effect of the residual stress and constituent material properties (fracture toughness of the matrix, Weibull modulus, and interfacial shear stress) on the bridging stress function was also considered in the numerical analysis. The results at 25 and 1400°C are given in Figs. 9 and 10.

Figure 9 shows the results of numerically calculated theoretical K_R-curves with the influence of residual stress of the fiber (σ_r^f) taken into consideration. Experimental results of fracture resistance at room temperature determined by *in-situ* crack length measurements are also shown in Fig. 9. The calculations for theoretical K_R-curves are made with the following values, σ_f = 2.8 GPa [19], m = 3, and τ = 6.3 MPa [7]. It has been shown in the previous study that fiber strength and Weibull modulus do not affect the K_R-curve significantly for a/W values in the range of 0.3 to 0.6 [7]. The residual stress in the fiber, however, can significantly affect the K_R-curve. The theoretical K_R values determined without taking residual stress of the fiber into account is higher than that determined with the residual stress for all a/W values. The K_R-curve determined at room temperature with residual fiber stress of -108.9 MPa (compression) [7] fits well with the experimental data up to $a/W \approx 0.7$. Beyond this point, the increase in the theoretical K_R-curve becomes larger than the experimental data due to the crack deflection and branching [7].

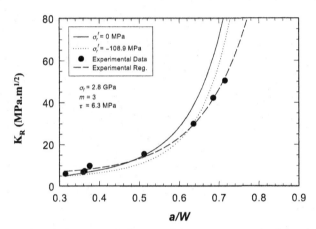

Fig. 9. Influence of residual stress and other materials parameters on the theoretical K_R behavior in SiC_f-Zircon composite at 25°C.

Figure 10 shows the results of the theoretical K_R-curves based on the numerical calculations performed at 1400°C. Experimental results of fracture resistance behavior at 1400°C are also shown in Fig. 10. The influence of a change in matrix toughness (K_m) on numerically calculated K_R-curves at 1400°C is also shown in Fig. 10. An interfacial shear stress of τ = 1 MPa is used to calculate the theoretical K_R-curve at this temperature, because of the reduced clamping force from residual stress at 1400°C. A decrease in the slope of the fracture resistance as a function of a/W at 1400°C is also observed with increasing toughness of the matrix from 5 to 10

MPa·m$^{1/2}$ as shown in the Fig. 10. These results indicate that the theoretical K_R-curve with the value of $K_m = 7$ MPa·m$^{1/2}$ and $\tau = 1$ MPa provided the best fit to the experimental K_R-curve for up to a/W < 0.6 at 1400°C.

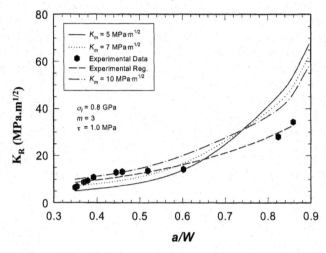

Fig. 10. Influence of residual stress and other materials parameters on the theoretical K_R behavior in SiC$_f$-Zircon composite at 1400°C.

These results also show a large difference between the experimental and theoretical K_R-curves at larger a/W (> 0.6) at 1400°C, which can be explained by multiple crack growth observed because of the significant creep deformation of the fiber and matrix phase above 1300°C. In this context, the experimental fracture resistance data may provide a lower bound of the fracture resistance if only the crack length of the dominant crack is used to determine the fracture resistance of the composites. Incorporation of the influence of the multiple cracking with the dominant crack is expected to increase the experimental fracture resistance curves, which will then be closer to the theoretical K_R-curve. These approaches will be pursued in future research works.

CONCLUSIONS

The fracture resistance behaviors of SiC$_f$-Zircon and SiC$_f$-SiC MI-composites were measured in the temperature range of 20°-1400°C. A novel *in situ* crack length measurement technique was employed to accurately measure the crack length as a function of the applied load. This study showed that both ceramic composites exhibited a rising *R*-curve behavior at room and elevated temperatures. Two different modes of crack growth were observed in SiC$_f$-Zircon composite . Below 1250°C, a single dominant crack was observed from the notch tip. At very high temperatures (1350°-1400°C), however, lower fracture resistance was obtained because of the weakening of the fiber and creep processes. Analytical model was developed and successfully used for predicting fracture resistance of CMCs over a range of temperatures.

ACKNOWLEDGEMENT
This research is partially supported by National Science Foundation through a Grant #CMS-0070213. Any opinions, findings, and conclusions or recommendations expressed in this material are those of the authors and do not necessarily reflect the views of the National Science Foundation.

REFERENCES
1. A. G. Evans, "Perspective on the Development of High-Toughness Ceramics," *Journal of the American Ceramic Society*, **73** [2] 187-206 (1990).
2. F. Zok and C. L. Hom, "Large Scale Bridging in Brittle Matrix Composites," *Acta Metallurgica et Materialia*, **38** [10] 1895-1904 (1990).
3. M. D. Thouless and A. G. Evans, "Effects of Pull-Out on the Mechanical Properties of Ceramic-Matrix Composites," *Acta Metallurgica*, **36** [3] 517-522 (1988).
4. B. N. Cox, D. B. Marshall, and M. D. Thouless, "Influence of Statistical Fiber Strength Distribution on Matrix Cracking in Fiber Composites," *Acta Metallurgica*, **37** [7] 1933-1943 (1989).
5. S. J. Bennison and B. R. Lawn, "Role of Interfacial Grain-Bridging Sliding Friction in the Crack-Resistance and Strength Properties of Nontransforming Ceramics," *Acta Metallurgica*, **37** [11] 2659-2671 (1989).
6. B. N. Cox, "Extrinsic Factors in the Mechanics of Bridged Cracks," *Acta Metallurgica et Materialia*, **39** [6] 1189-1201 (1991).
7. Y.-L.Wang, U. Anandakumar, and R. N. Singh, "Effect of Fiber Bridging Stress on the Fracture Resistance of SiC$_f$/Zircon Ceramics Composites," *Journal of the American Ceramic Society*, 83[3], 1207-1214 (2000).
8. U. Anandakumar and R. N. Singh, "Creep Response of Fiber Reinforced Ceramic Composites," *Ceramic Engineering and Science Proceedings*, **20** [4] 451-458 (1999).
9. K.L. Luthra, R.N. Singh, and M. Brun, " Toughened Silcomp Composites," *Amer. Ceram. Soc. Bull.*, **72**[7], 79-85 (1993).
10. Standard Test Method for Plane-Strain Fracture Toughness of Metallic Materials, *ASTM Standard*, E399-83.
11. J. LLorca and M. Elices, "Fracture Resistance in Fiber-Reinforced Ceramic Matrix Composites," *Acta Metall. Mater.*, **38** [12] 2485–2492 (1990).
12. J. LLorca and M. Elices, "A Simplified Model to Study Fracture Behavior in Cohesive Materials," *Cem. Concr. Res.*, **20**, 92–102 (1990).
13. J. LLorca and R. N. Singh, "Influence on Fiber and Interfacial Properties on Fracture Behavior of Fiber-Reinforced Ceramic Composites," *J. Am. Ceram. Soc.*, **74** [11] 2882–2890 (1991).
14. B. N. Cox, D. B. Marshall, and M. D. Thouless, "Influence of Statistical Fiber Strength Distribution on Matrix Cracking in Fiber Composites," *Acta Metall.*, **37** [7] 1933–1943 (1989).
15. D. B. Marshall and B. N. Cox, "The Mechanics of Matrix Cracking in Brittle-Matrix Fiber

Composites," *Acta Mater.*, **33** [11] 2013–2021 (1985).

16. R. M. L. Foote, Y.W. Mai, and B. Cotterell, "Crack Growth Resistance Curves in Strain-Softening Materials," *J. Mech. Phys. Solids,* **34**, 593–607 (1986).

17. C.-H. Hsueh and P. F. Becher, "Evaluation of Bridging Stress from R-Curve Behavior for Nontransforming Ceramics," *J. Am. Ceram. Soc.,* **71** [5] C-234–C-237 (1988).

18. R. W. Steinbrech, A. Reichl, and W. Schaarwächter, "R-Curve Behavior of a Long Cracks in Alumina," *J. Am. Ceram. Soc.,* **73** [7] 2009–2015 (1990).

19. D. B. Marshall and A. G. Evans, "The Influence of Residual Stress on the Toughness of Reinforced Brittle Materials," *Materials Forum*, **11**, 304 (1988).

SHEAR STRENGTH BEHAVIORS OF CERAMIC MATRIX COMPOSITES AT ELEVATED TEMPERATURES

Sung R. Choi,[*] Narottam P. Bansal, Anthony M. Calomino, and Michael J. Verrilli
National Aeronautics & Space Administration, Glenn Research Center, Cleveland, OH 44135

ABSTRACT

Interlaminar shear strength of four different fiber-reinforced ceramic matrix composites was determined with double-notch shear test specimens as a function of test rate at elevated temperatures ranging from 1100 to 1316 °C in air. Shear strength degradation with decreasing test rate was significant for SiC/MAS and C/SiC composites, but insignificant for SiC/SiC and Sylramic SiC/SiC composites. A phenomenological, power-law delayed failure model was proposed to account for and to quantify the rate dependency of shear strength of the composites. Additional stress rupture testing was conducted to validate the proposed model. The model was in good agreement with SiC/MAS and C/SiC composites, but in poor to reasonable agreement with Sylramic SiC/SiC. Constant shear stress-rate testing was proposed as a possible means of life prediction testing methodology for composites in shear when short lifetimes are expected.

INTRODUCTION

The successful development and design of continuous fiber-reinforced ceramic matrix composites (CMCs) are dependent on understanding their basic properties such as deformation, fracture, and delayed failure (fatigue, slow crack growth, or damage accumulation) behavior. Particularly, accurate evaluation of delayed failure behavior under specified loading/environment conditions is prerequisite to ensure accurate life prediction of structural components at elevated temperatures.

Although fiber-reinforced CMCs have shown improved resistance to fracture and increased damage tolerance compared with the monolithic ceramics, inherent material/processing defects or cracks in the matrix-rich interlaminar regions can still cause delamination under interlaminar normal or shear stress, resulting in loss of stiffness or in some cases structural failure. Strength behavior of CMCs in shear has been characterized in view of their unique interfacial architectures and its importance in structural applications [1-4]. Because of the inherent nature of *ceramic* matrix composites, it would be highly feasible that interlaminar defects or cracks are susceptible to delayed failure particularly at elevated temperatures, resulting in strength degradation or time-dependent failure. Although delayed failure is one of the important life-limiting phenomena, few studies have been done on this subject for CMCs under *shear* at elevated temperatures.

In a previous study [5], both interlaminar and in-plane shear strengths of a unidirectional Hi-Nicalon™ fiber-reinforced barium strontium aluminosilicate (SiC/BSAS) composite were determined at 1100 °C in air as a function of test rate using double-notch shear test specimens. The composite exhibited a significant effect of test rate on shear strength, regardless of orientation. The shear strength degraded by about 50% as test rate decreased from the highest

---

[*] Corresponding author; NASA Resident Principal Scientist, Ohio Aerospace Institute; Tel.: 216-433-8366; fax:216-433-8300.
Email addresss: sung.r.choi@grc.nasa.gov.

(10^2 MPa/s) to the lowest (10^{-2} MPa/s). A phenomenological, life-prediction model has been proposed and formulated to account for the shear strength degradation of the composite.

This paper, as an extension of the previous study, describes shear strength behavior of four different fiber-reinforced CMCs at elevated temperatures, including three SiC fiber-reinforced CMCs and one carbon-fiber reinforced CMC. Interlaminar shear strength of each composite was determined in double notch shear as a function of test rate using constant stress-rate testing. The shear strength behavior was analyzed using the power-law type of model proposed previously in order to quantify the rate dependency and delayed failure of the composites under shear. Stress rupture testing in shear was also conducted with three chosen CMCs in an attempt to validate the proposed model.

EXPERIMENTAL PROCEDURES

Four different CMCs—three SiC fiber-reinforced and one carbon fiber-reinforced—were used in this study, including Nicalon™ SiC crossplied (2–D) fiber-reinforced magnesium aluminosilicate (designated SiC/MAS), Nicalon™ SiC plain-woven (2–D) silicon carbide (designated SiC/SiC), Sylramic SiC plain woven (2-D) silicon carbide (designated Sylramic SiC/SiC), and T-300 carbon-fiber plain-woven (2–D) silicon carbide (designated C/SiC).

The SiC/MAS composites were fabricated through hot pressing followed by ceraming of the composites by a thermal process. The silicon carbide matrix in the SiC/SiC composite was processed through chemical vapor infiltration (CVI) into the fiber preforms. More detailed information regarding the processing of these SiC/MAS and SiC/SiC composites can be found elsewhere [6]. The Sylramic cloth preforms in the Sylramic SiC/SiC composite were stacked and chemically vapor infiltrated with a thin BN-based interface coating followed by SiC matrix over-coating. Remaining matrix porosity was filled with SiC particulates and then with molten silicon at 1400°C, a process termed slurry casting and melt infiltration [7]. The carbon fiber performs in the C/SiC composite were coated with pyrolytic carbon as an interface prior to CVI SiC infiltration [8]. Information regarding fiber volume fraction and laminate architecture of the test composites is summarized in Table 1.

The double-notch-shear (DNS) test specimens were machined from each composite laminate. Typically, test specimens were 13-15 mm wide (W) and 30 mm long (L). The thickness of test specimens was the same as a nominal thickness of each laminate (see Table 1). Two notches, 0.3 mm wide (h) and 6 mm (L_n) away from each other, were made into each test specimen such that the two notches were extended to the middle of each specimen within ±0.05 mm so that shear failure occurred on the plane between the notch tips. Schematics of DNS test specimen showing a notch configuration is shown in Figure 1. Monotonic shear testing for DNS test specimens was conducted at elevated temperatures with different test rates in ambient air (relative

Table 1. Fiber-reinforced ceramic matrix composites used in this work

Material	Architecture	Fiber	Fiber volume fraction	Laminates^	Manufacturer
SiC/MAS	2-D c-plied	Nicalon# SiC	0.39	HP; 16 plies; 3 mm thick	Corning
SiC/SiC	2-D woven	Nicalon SiC	0.39	CVI; 12 plies; 3.5 mm thick	E. I. Du Pont
Sylramic SiC/SiC	2-D woven	Sylramic* SiC	0.36	CVI, SC,MI; 8 plies; 2 mm thick	GE Power S. Comp.
C/SiC	2-D woven	Carbon (T300)	0.46	CVI; 26 plies; 3.3 mm thick	Honeywell Adv. Comp.

Nippon Carbon (Japan); * Dow Corning (MI)
^ HP: hot pressed; CVI: chemical vapor infiltration; SC: slurry casting; MI: melt infiltration

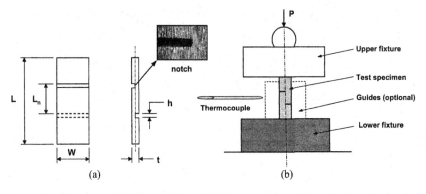

(a)   (b)

Figure 1. Configurations of double-notch shear (DNS) test specimen (a); and schematic showing test fixture and test specimen used in this work (b).

humidity of about 45%), using an electromechanical test frame (Model 8562, Instron, Canton, MA) under load control. This testing, employing with different test rates, is called constant stress-rate or 'dynamic fatigue' testing that is used in brittle materials to estimate their time-dependent or slow crack growth behavior [9,10]. Test temperatures were 1100, 1200, 1316, and 1200 °C, respectively, for the SiC/MAS, SiC/SiC, S-SiC/SiC, and C/SiC composites.

A total of three to four test rates ranging from 10^{-4} to 10^1 MPa/s were used for a given composite, depending on the type of composite. Typically, three to ten test specimens were tested at each test rate, again depending on type and availability of materials. A simple test-fixture configuration consisting of SiC upper and lower fixtures, as shown in Figure 1, was used for test specimens whose thickness was ≥ 3 mm. With this specimen thickness and the tight machining tolerances, the test specimens could stand alone and be subjected to negligible bending (≤ 4 %) due to misalignment, geometrical inaccuracies, and/or buckling. By contrast, for thin test specimens whose thickness was about 2 mm (e.g., S-SiC/SiC composite), anti-buckling guides were used in conjunction with specially designed ring-shaped fixtures. Each test specimen was held for about 20 min at the test temperature for thermal equilibration prior to testing. Test specimen configurations and testing procedures were followed in accordance with ASTM test method C 1425 [11]. The shear fracture stress—the average shear stress at failure—was calculated using the following relation

$$\tau_f = \frac{P_f}{WL_n} \tag{1}$$

where τ_f is the shear strength, P_f is the fracture force, and W and L_n are the specimen width and the distance between the two notches, respectively (see Figure 1).

Additionally, stress rupture testing in shear was conducted with DNS test specimens for SiC/MAS at 1100°C, S-SiC/SiC at 1316 °C, and C/SiC at 1200 °C in air. The test frame, test specimen configuration, and test fixture used in stress rupture were the same as those used in

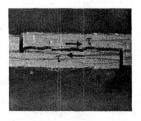

Figure 2. A typical example showing interlaminar shear failure in double-notch shear testing for Sylramic SiC/SiC composite at 1316 °C in air. A test specimen prior to testing (at right) is included for comparison.

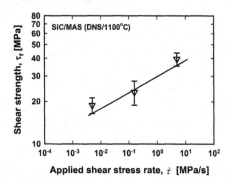

Figure 3. Results of interlaminar shear strength as a function of applied shear test rate for 2-D crossplied SiC/MAS composite tested at 1100 °C in air. The solid line represents the best-fit.

constant stress-rate testing. For each composite, the number of applied stresses was typically three, and the number of test specimens was six. This stress rupture testing was performed to determine delayed failure behavior under constant applied stress and to validate the phenomenological life prediction model.

EXPERIMENTAL REULTS

Constant Stress-Rate Testing

Without exception, all specimens tested failed in shear mode along their respective shear planes. A typical example of a tested specimen showing such shear mode failure is presented in Figure 2, together with a test specimen prior to testing for comparison.

SiC/MAS composite: The results of monotonic shear strength testing for the 2-D crossplied SiC/MAS composite tested at 1100 °C are presented in Figure 3, where *shear strength* is plotted as a function of *applied test rate*. The solid line in the figure represents a best-fit regression based on the log (*shear strength*) versus log (*applied test rate*) relation (the reason for using the log-log relation will be described in the "Discussion" section.). The decrease in shear strength with decreasing test rate, indicating a susceptibility to delayed failure, was significant for this

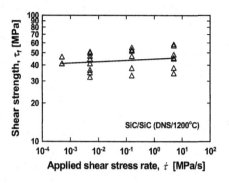

Figure 4. Results of interlaminar shear strength as a function of applied shear test rate for 2-D woven SiC/SiC composite tested at 1200 °C in air. The solid line represents the best-fit.

composite. The shear strength degradation was about 50% when test rate decreased from the highest (5 MPa/s) to the lowest (0.005 MPa/s) value. A similar trend in shear strength degradation with decreasing test rate was also found from a previous study in a 1-D, Hi-Nicalon fiber-reinforced barium strontium aluminosilicate (SiC/BSAS) composite at 1100 °C in air [5]. This trend in shear strength with respect to test rate was also analogous to that in ultimate tensile strength of various continuous fiber-reinforced CMCs including SiC/MAS, SiC/CAS (calcium aluminosilicate), SiC/BSAS, C/SiC, and SiC/SiC composites [12,13]. These CMCs have exhibited significant degradation of ultimate tensile strength with decreasing test rates, with their degree of degradation being dependent on material and test temperature.

Fracture surfaces of the SiC/MAS composite showed that the mode of shear failure was typified as delamination of fibers from matrix-rich regions, implying that the fiber-matrix interfacial architecture is the most influencing characteristic in controlling shear properties of the composite. The presence of viscous flow/phases was obvious from fracture surfaces, particularly at low test rates in which more enhanced delayed failure occurred. The residual glassy phase might have been a major cause of delayed failure in the SiC/MAS composite, as observed in the SiC/BSAS composite [5].

SiC/SiC Composite: Figure 4 shows the results for the 2-D woven SiC/SiC composite tested at 1200 °C, in which shear strength is plotted in the same way as in Figure 3, with the line representing the best-fit. Unlike the SiC/MAS composite, the SiC/SiC composite did not exhibit any significant strength degradation with decreasing test rate. The strength degradation from the highest (≈5 MPa/s) to the lowest (0.0005 MPa/s) test rate was only 5%, indicating a greater resistance to delayed failure, compared with the SiC/MAS composite. Fracture surfaces showed that some oxidation occurred therein with a bluish discoloration. Of course, more discoloration at lower test rates, and vice versa. Fracture surfaces also exhibited that well developed shear occurred between two adjacent laminates by interfacial delamination so that damage in fiber tows observed was insignificant. Also, except for discoloration, the difference in fracture surface of test specimens between high and low test rates appeared to be unnoticeable.

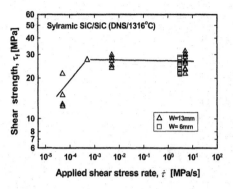

Figure 5. Results of interlaminar shear strength as a function of applied shear test rate for 2-D woven Sylramic SiC/SiC composite tested at 1316 °C in air. The solid line represents the best-fit.

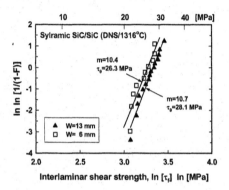

Figure 6. Weibull shear strength distributions of two different sizes (W=13 and 6 mm) of double-notch shear test specimens of Sylramic SiC/SiC composite tested at 5 MPa/s at 1316 °C in air. m:Weibull modulus; τ_θ: characteristic shear strength.

Sylramic SiC/SiC Composite: The results of shear strength testing for the Sylramic fiber-reinforced SiC (Sylramic SiC/SiC) composite at 1316 °C are depicted in Figure 5. The composite exhibited no apparent strength degradation down to 0.0005 MPa/s; however, it exhibited strength degradation at the lowest test rate of 0.00005 MPa/s, resulting in an about 40 % reduction in strength. The test time corresponding to the lowest test rate was around 85 h so that this long duration of test would have caused such significant strength degradation through a possible environmental degradation. Invariably, a severe discoloration was observed from the fracture surfaces of specimens tested at this test rate. Consequently, a question arises as to whether strength degradation would take place for the SiC/SiC composites if an extremely low test rate of 0.00005 MPa/s is applied to the SiC/SiC composite, which requires a further study. The pattern of shear fracture of this 2-D woven Sylramic SiC/SiC composite was very similar to

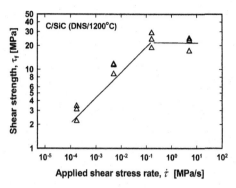

Figure 7. Results of interlaminar shear strength as a function of applied shear test rate for 2-D woven C/SiC composite tested at 1200 °C in air. The solid line represents the best-fit.

that of the 2-D woven SiC/SiC counterpart: an apparent shear delamination with little damage in fiber tows was evident from fracture surfaces.

It is important to note that the size effect on shear strength was negligible for this composite, as seen in Figure 5, where interlaminar shear strength of additional test specimens with W=6 mm, determined at the same test rate of 5 MPa/s, was compared with that of the regular test specimens with W=13 mm. The average shear strength of 10 test specimens was 27±4 and 25±3 MPa, respectively, for W=13 and 6 mm. Also, considering that intrelaminar shear strength was almost the same regardless of test rate >0.00005 MPa/s, it could be possible to take all strength values at different test rates as a single representative data pool. With this in mind, Weibull two-parameter strength distributions were made for the two specimen widths and the result are shown in Figure 6. Weibull modulus (m=10.7 vs. 10.4) as well as characteristic shear strength (τ_θ =28 vs. 26 MPa) is about the same for both specimen sizes. It is noted that the value of Weibull modulus in shear seemed to be low as compared with that in tension in which Weibull modulus is typically around $m \approx 20$.

C/SiC Composites: Figure 7 shows the results of interlaminar shear strength testing for the C/SiC composite. Strength degradation between test rates of 5 and 0.16 MPa/s was negligible. However, below 0.16 MPa/s, strength degradation with decreasing test rate was very significant with a resulting degradation of about 90% when test rate decreased from 0.16 to 0.00017 MPa/s. The composite, in fact, was losing almost its shear-load bearing capability at that lowest test rate with shear strength of only τ_f= 2.9±0.6 MPa, compared with τ_f = 22-24 MPa at higher test rates.

Severe oxidation was observed to have occurred throughout the material body particularly at lower test rates. Powdered residues were often found from the fracture surfaces of these test specimens, an evidence of oxidation products. Also, weight loss of test specimens before and after tests occurred: the more weight loss at the lower test rate and vice versa. Figure 8 shows weight loss as a function of test time. The figure also included the weight loss data obtained from stress rupture tests. Regardless of the type of testing, weight loss increased exponentially with test time and reached a plateau around a test time of about 4 h. The corresponding weight loss at this time was about 40%, which is about 80% of the original fiber weight fraction. Hence,

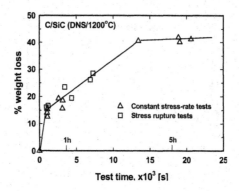

Figure 8. Weight loss as a function of test time in constant stress-rate and stress rupture testing for 2-D woven C/SiC composite tested at 1200 °C in air.

strength degradation of the C/SiC composite was due to oxidation of fibers so that effective number of fibers decreased with time, resulting in a decrease in load carrying capability. Evidence of oxidation during testing was also viewed from the nonlinearity of load vs. displacement curves. Other composites did not exhibit any measurable weight loss by oxidation and showed little nonlinearity in the stress-strain curves associated with creep deformation as well.

DISCUSSION

Model

It has been shown that strength degradation in tension at elevated temperatures with decreasing test rate occurred not only in CMCs such as SiC/CAS, SiC/MAS, SiC/SiC, C/SiC, SiC/BSAS (2-D) [12,13] but also in advanced monolithic ceramics including silicon nitrides, silicon carbides, and aluminas [14]. The strength degradation is known as a delayed failure (slow crack growth, SCG) phenomenon, commonly formulated by the empirical power-law relation [15]

$$v = \alpha(K_I / K_{Ic})^n \tag{2}$$

where v, K_I, and K_{Ic} are crack growth rate, stress intensity factor, and fracture toughness under Mode I loading, respectively. α and n are called delayed fatigue (or SCG) parameters. Based on this power-law relation, the tensile strength (σ_f) can be derived as a function of applied test rate or stress rate ($\dot{\sigma}$) with some mathematical manipulations [9,10,16]

$$\log \sigma_f = \frac{1}{n+1} \log \dot{\sigma} + \log D \tag{3}$$

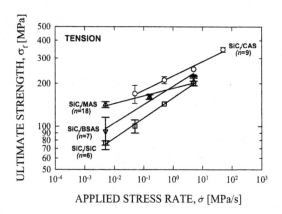

APPLIED STRESS RATE, $\dot{\sigma}$ [Mpa/s]

Figure 9. Examples of ultimate tensile strength as a function of applied stress rate at elevated temperatures in air for various CMCs [12,13] including SiC/MAS (2-D crossplied; 1100 °C), SiC/CAS (1-D; 1100 °C), SiC/SiC (2-D woven; 1200 °C), and SiC/BSAS (2-D crossplied; 1100 °C).

where D is another SCG parameter dependent on inert strength, n, and crack geometry.

A test methodology based on Eq. (3) is called constant stress-rate ("dynamic fatigue") testing and has been established as ASTM test methods C1368 [9] and C1465 [10] to determine SCG parameters of advanced monolithic ceramics at ambient and elevated temperatures. It has been shown that Eq. (3) was applicable even to CMCs including SiC/CAS (1-D), SiC/MAS (2-D), SiC/SiC (2-D), C/SiC (2-D), and SiC/BSAS (2-D) tested in tension at 1100 to 1200 °C in air, indicating that delayed failure of those composites would be adequately described by the power-law formulation, Eq. (2). Examples of some CMCs showing invariably significant strength degradation and good data fit to Eq. (3) [12,13] are presented in Figure 9.

It has been also observed that 1-D SiC/BSAS composite exhibited significant strength degradation in shear at 1100 °C in air [5]. A phenomenological model in shear has been proposed and was found to describe well the shear strength dependency of the composite on test rate. The proposed model, similar in expression to the power-law relation of Eq. (2), takes the following fundamental delayed failure formulation [5]:

$$v_s = \frac{da}{dt} = \alpha_s (K_{II} / K_{IIc})^{n_s} \qquad (4)$$

where v_s, a, t, K_{II}, and K_{IIc} are the crack growth rate in shear, crack size, time, Mode II stress intensity factor, and Mode II fracture toughness, respectively. α_s and n_s are delayed failure parameters in shear. In monotonic shear testing, a constant displacement rate or constant force rate is applied to a test specimen until the test specimen fails, so that the shear stress applied to the test specimen is a linear function of test time:

$$\tau = \int_0^t \dot{\tau}(t)dt = \dot{\tau}\,t \tag{5}$$

where τ is (remote) shear stress, and $\dot{\tau}$ is applied shear stress rate. The applied shear stress rate $\dot{\tau}$ in displacement control can be determined from the slope $(\Delta P/\Delta t)$ of each recorded force-versus-time curve including the portion at or near the point of fracture but excluding the initial nonlinear portion, if any, using a relation $\dot{\tau} = (\Delta P/\Delta t)[1/(WL_n)]$. In case of load control, the applied shear stress rate can be obtained, without determining the slope, directly from a relation $\dot{\tau} = \dot{P}/(WL_n)$, where $\dot{P}$ is the applied force rate employed within a test frame. The generalized expression of stress intensity factor in shear takes the following form [17]:

$$K_{II} = Y_s\tau\,a^{1/2} \tag{6}$$

where Y_s is a crack geometry factor in shear. Using Eqs. (4), (5), and (6) and following a similar procedure as used to derive Eq. (3) in Mode I, one can obtain shear strength (τ_f) as a function of applied shear stress rate as follows:

$$\log\tau_f = \frac{1}{n_s+1}\log\dot{\tau} + \log D_s \tag{7}$$

where $D_s = [B_s(n_s+1)\,\tau_i^{n_s-2}]^{1/n_s+1}$ with $B_s = 2K_{Ic}/[\alpha_s Y_s^2(n_s-2)]$ and τ_i being the inert shear strength. Equation (7) is identical in form to Eq. (3) for Mode I loading. Delayed failure parameters n_s and D_s in shear can be determined from slope and intercept of a linear regression analysis of the log (*individual shear strength with units of MPa*) vs. log (*individual shear stress rate with units of MPa/s*) data, based on Eq. (7). The parameter α_s can be estimated from the D_s expression with appropriate constants and parameters.

A summary of interlaminar shear strength as a function of test rate, based on Eq. (7), for the composites used in this study is shown in Figure 10. The level of shear strength and the degree of strength degradation with respect to test rate are clearly quantified. The figure also includes delayed failure parameters n_s for each composite and the previous result for the SiC/BSAS composite [5]. Again, the reasonable data fit to Eq. (7) indicates that the fundamental power-law formulation of Eq. (4) can describe very reasonably the delayed failure behavior of the current composite materials in shear.

Table 2 is a summary of the parameters n_s and D_s and the coefficients of correlation (r_{coef}) in curve fitting for all the composites. The table also includes the parameters n and D in tension [12,13]. Aluminosilicate CMCs and C/SiC exhibit a significant susceptibility to delayed failure both in shear and tension with $n_s<12$ and $n\leq13$, while SiC/SiC exhibit an insignificant susceptibility to delayed failure in shear with $n_s>90$ and a moderate susceptibility in shear with $n=26$. The reason for the appreciable delayed-failure susceptibility in either shear or tension for aluminosilicate CMCs may be attributed to viscous flow by residual glassy phase occurring at interfaces. However, it must be noted that ultimate tensile strength of a composite is mainly governed by fibers; whereas, shear strength is more likely controlled by interface architectures.

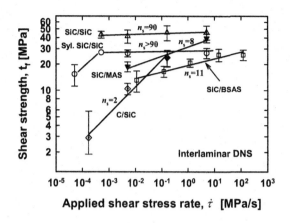

Applied shear stress rate, $\dot{\tau}$ [MPa/s]

Figure 10. Summary of interlaminar shear strength as a function of applied shear stress rate for SiC/MAS (2-D crossplied) at 1100 °C, SiC/SiC (2-D woven) at 1200 °C, Sylramic SiC/SiC (2-D woven) at 1316 °C, and C/SiC (2-D woven) at 1200 °C. The interlaminar shear data on SiC/BSAS (1-D) at 1100 °C [5] is also included.

Table 2. Delayed failure parameters of n_s, D_s, and cefficient of correlation (r_{coef}) in shear for various CMCs determined in constant stress-rate testing at elevated temperatures. Values of n and D in tension were also included for comparison

Composites	Test temperature (°C)	In shear			In tension [12,13]	
		n_s	D_s	r_{coef}	n	D
SiC/MAS (2-D)	1100	8.3	31.8	0.9283	18	185
SiC/SiC (2-D)	1200	90	44.8	0.6693	6	158
Sylramic SiC/SiC (2-D)	1316	>90 (3)*	27	-	-	-
C/SiC (2-D)	1200	2.3	44.7	0.9721	6	196
SiC/BSAS (1-D) [6]	1100	11.2	19.2	0.8442	7	188

* The value of n_s in the parenthesis for Sylramic SiC/SiC composite was estimated based on the two lowest test rates of 0.00005 and 0.0005 MPa/s (see also Figure 5).

Verification

Constant stress-rate testing in tension has been shown to be a possible alternative to life prediction testing, as verified with stress rupture testing for various CMCs at elevated temperatures of 1100 to 1200 °C [12,13]. This indicated that the same failure mechanism might have been operative, independent of loading configuration that was either in monotonic (constant stress-rate) loading or in static (stress rupture) loading. In the same way, it would be expected that life prediction in *shear* from one loading configuration to another could be made analytically or numerically as far as a single delayed-failure mechanism would also be dominant in shear, regardless of loading configuration. A phenomenological, simplified life prediction approach was proposed based on the data of n_s and D_s using the following relation that accounts for shear loading (modified from a relation primarily used for brittle materials in tension) [5]:

$$t_f = [\frac{D_s^{n_s+1}}{n_s+1}] \tau^{-n_s} \tag{8}$$

where t_f is time to failure and, τ is applied shear stress. Equation (8) determines the life for a given applied constant shear stress. Statistically, the prediction represents the time to failure at a failure probability of approximately 50%.

In order to verify the life prediction relation proposed in Eq. (8), stress rupture testing, aforementioned in the Experimental Procedures section, was conducted for the SiC/MAS, Sylramic SiC/SiC, and C/SiC composites. The results of stress rupture testing are presented in Figure 11. Significant delayed failure in shear occurred in the SiC/MAS and C/SiC composites. Life prediction was made using the constant stress-rate data n_s and D_s for each composite, based on the proposed relation, Eq. (8), and the resulting prediction is presented as a solid line in the figure. Despite a limited number of test specimens used, the prediction for the SiC/MAS and C/SiC composite was in good agreement with the experimental data determined in stress rupture, thereby validating Eq. (8). The prediction for the Sylramic SiC/SiC composite, however, was in poor to reasonable agreement. Note that scatter of time to failure in stress rupture testing is considerably greater than that of strength in monotonic (constant stress-rate) testing for many brittle materials. Therefore, a wide range of materials with a greater number of test specimens, particularly in stress rupture testing, is needed to rigorously verify the proposed model.

Unlike the other CMCs, the C/SiC composite was subjected to significant oxidation of carbon fibers, resulting in material loss. Therefore, strength degradation was increased, attributed to decreasing fiber volume fraction and subsequently increasing porosity. The stress-oxidation model proposed previously [18] could give a better physical explanation regarding the strength degradation/time dependency phenomenon. However, the phenomenological power-law formulation used in this study still provides a simple, convenient way to quantify the degree of strength degradation with respect to test rate. The oxidation-induced damage would be considered to be equivalent to crack-like flaws growing through matrix-fiber interfaces from a fracture-mechanics point of view. The equivalent crack would propagate under a driving force (K_{II}) based on Eq. (4) so that the resulting strength follows Eq. (7). The fact that there was good agreement between the prediction and the stress rupture data implies that the fundamental driving force formulated by Eq. (4) would have controlled delayed shear failure of the C/SiC composite, irrespective of loading configurations. Stress/life behavior of the C/SiC composite has been extensively studied in a low partial pressure of oxygen at elevated temperatures [8,19].

The results of interlaminar-shear strength behavior of various CMCs showed that constant stress-rate testing could be applicable to determine phenomenological life prediction parameters of the composites even in shear. This was evidenced by the results of constant stress-rate testing and in part by the results of additional stress rupture testing for some chosen CMCs, indicating that the overall failure mechanism in shear could be the one governed by the power-law type of delayed failure. The merit of constant stress rate testing along with the power-law formulation

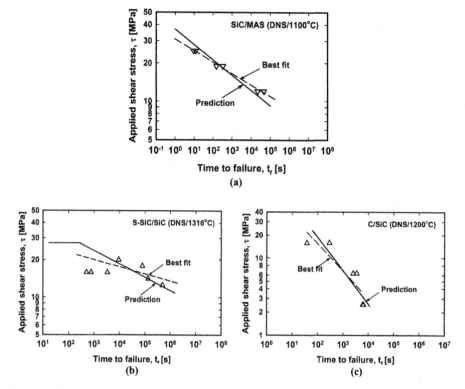

Figure 11. Results of stress rupture testing in shear for: (a) SiC/MAS at 1100 °C; (b) Sylramic SiC/SiC at 1316 °C; (c) C/SiC at 1200 °C. Life prediction based on Eq. (8) is shown as a solid line for each composite. Best-fit line (log τ vs. log t_f) is also included for each composite.

is enormous in terms of simplicity, test economy (short test times), and less data scatter over other stress rupture or cyclic fatigue testing. Although the experimental results and phenomenological model are presented in this work, a more detailed study of the shear failure mechanism regarding the microscopic influences, which include matrix/fiber interaction, matrix cracking, and environmental effects [20-22], is still needed. It should be kept in mind that the phenomenological model proposed here may incorporate other operative models such as viscous sliding, void nucleation, and coalescence, etc., which can be all covered under a generic term of delayed failure, slow crack growth, fatigue, or damage initiation/accumulation.

Finally, the results of this work also suggest that care must be exercised when characterizing elevated-temperature shear strength of composite materials. This is due to the fact that if a material exhibits rate dependency, elevated-temperature shear strengths are relative: the shear strength simply depends on which test rate one chooses. Therefore, at least two test rates (high and low) are recommended to better characterize high-temperature shear strength behavior of a composite material, as suggested previously for the determination of ultimate tensile strength of CMCs at elevated temperatures [12,13].

CONCLUSIONS

Interlaminar shear strength of four different CMCs including SiC/MAS, SiC/SiC, Sylramic SiC/SiC, and C/SiC was determined using different test rates at temperatures ranging from 1100 to 1316 °C in air. Shear strength degradation in terms of decreasing test rate was significant for SiC/MAS and C/SiC composites and was insignificant for SiC/SiC and Sylramic SiC/SiC composites. The Sylramic SiC/SiC composite, however, exhibited some degree of strength degradation at a very low test rate of 0.00005 MPa/s. A phenomenological, power-law delayed failure model was used to account for strength degradation and showed good agreement with SiC/MAS and C/SiC but in poor to reasonable agreement with Sylramic SiC/SiC, as compared with the additional stress rupture results. Constant-shear stress-rate testing could be a possible means of life prediction test methodology for composites in shear in case that short lifetimes of components are expected.

Acknowledgements

This work was supported in part by the Ultra-Efficient Engine Technology (UEET) Project, NASA Glenn Research Center, Cleveland, Ohio. The authors are grateful to R. Pawlik for experimental work during the course of this work.

REFERENCES

1. P. Brondsted, F. E. Heredia, and A. G. Evans, "In-Plane Shear Properties of 2-D Ceramic Composites," *J. Am. Ceram. Soc.*, **77**[10] 2569-2574 (1994).
2. E. Lara-Curzio and M. K. Ferber, "Shear Strength of Continuous Fiber Ceramic Composites," ASTM STP 1309, p. 31, American Society for Testing & Material, West Conshohocken, PA (1997).
3. N. J. J. Fang and T. W. Chou, "Characterization of Interlaminar Shear Strength of Ceramic Matrix Composites," *J. Am. Ceram. Soc.*, **76**[10] 2539-2548 (1993).
4. Ö. Ünal and N. P. Bansal, "In-Plane and Interlaminar Shear Strength of a Unidirectional Hi-Nicalon Fiber-Reinforced Celsian Matrix Composite," *Ceramics International*, **28** 527-540 (2002).
5. S. R. Choi and N. P. Bansal, "Shear Strength as a Function of Test Rate for SiC$_f$/BSAS Ceramic Matrix Composite at Elevated Temperature," accepted in *J. Am. Ceram. Soc.*, (2004).
6. D. W. Worthem, "Thermomechanical Fatigue Behavior of Three CFCCs," NASA CR-195441, National Aeronautics & Space Administration, Glenn Research Center, Cleveland, OH (1995).
7. D. Brewer, "HSR/EPM Combustor Materials Development Program," *Mat. Sci. Eng.*, **A261** 284-291 (1999).
8. M. J. Verrilli, A. Calomino, and D. J. Thomas, "Stress/Life Behavior of a C/SiC Composite in a Low Partial Pressure of Oxygen Environment: I- Static Strength and Stress Rupture Database," *Ceram. Eng. Sci. Proc.*, **23**[3] 435-442 (2002).
9. ASTM C 1368, "Standard Test Method for Determination of Slow Crack Growth Parameters of Advanced Ceramics by Constant Stress-Rate Flexural Testing at Ambient Temperature,"

Annual Book of ASTM Standards, Vol. 15.01, American Society for Testing and Materials, West Conshohocken, PA (2003).

10. ASTM C 1465, "Standard Test Method for Determination of Slow Crack Growth Parameters of Advanced Ceramics by Constant Stress-Rate Flexural Testing at Elevated Temperatures," *Annual Book of ASTM Standards*, Vol. 15.01, American Society for Testing and Materials, West Conshohocken, PA (2003).

11. ASTM C 1425, "Test Method for Interlaminar Shear Strength of 1-D and 2-D Continuous Fiber-reinforced Advanced Ceramics at Elevated Temperatures," *Annual Book of ASTM Standards*, Vol.15.01, American Society for Testing & Materials, West Conshohocken, PA (2002).

12. S. R. Choi and J. P. Gyekenyesi, "Effect of Load Rate on Tensile Strength of Various CFCCs at Elevated Temperatures: An Approach to Life-Prediction Testing," *Ceram. Eng. Sci., Proc.*, **22**[3] 597-606 (2001).

13. S. R. Choi, N. P. Bansal, and M. J. Verrilli, "Delayed Failure of Ceramic Matrix Composites in Tension at Elevated Temperatures," in press *J. Euro. Ceram. Soc.* (2004).

14. S. R. Choi and J. P. Gyekenyesi, "'Ultra'-Fast Fracture Strength of Advanced Structural Ceramics at Elevated Temperatures: An Approach to High-Temperature 'Inert' Strength," pp. 27-46 in *Fracture Mechanics of Ceramics*, Vol. 13, Edited by R. C. Bradt, D. Munz, M. Sakai, V. Ya. Shevchenko, and K. W. White, Kluwer Academic/Plenum Publishers, New York (2002).

15. S. M. Wiederhorn, "Subcritical Crack Growth in Ceramics," pp. 613-646 in *Fracture Mechanics of Ceramics*, Vol. 2, Edited by R. C. Bradt, D. P. H. Hasselman, and F. F. Lange, Plenum Press, New York (1974)

16. A. G. Evans, Slow Crack Growth in Brittle Materials under Dynamic Loading Condition," *Int. J. Fracture*, **10** 251-259 (1974).

17. G. C. Sih, Handbook of Stress Intensity Factors, p. 3.2.1-2, Lehigh University, Bethlehem, PA (1973).

18. M. C. Halbig, "Modeling the Oxidation of Carbon Fibers in a C/SiC Composite under Stressed Oxidation," Ceram. Eng. Sci. Proc., **23**[3] 427-434 (2002).

19. A. Calomino, M. J. Verrilli, and D. J. Thomas, "Stress/Life Behavior of a C/SiC Composite in a Low Partial Pressure of Oxygen Environment: II- Stress Rupture Life and Residual Strength Relationship," *Ceram. Eng. Sci. Proc.*, **23**[3] 443-451 (2002).

20. C. A. Lewinsohn, C. H. Henager and R. H. Jones, "Environmentally Induced Time-Dependent Failure Mechanism in CFCCS at Elevated Temperatures," *Ceram. Eng. Sic. Proc.*, **19**[4] 11-18 (1998).

21. C. H. Henager and R. H. Jones, "Subcritical Crack Growth in CVI Silicon Carbide Reinforced with Nicalon Fibers: Experiment and Model," *J. Am. Ceram. Soc.*, **77**[9] 2381-94 (1994).

22. S. M. Spearing, F. W. Zok and A. G. Evans, "Stress Corrosion Cracking in a Unidirectional Ceramic-Matrix Composite, *J. Am. Ceram. Soc.*, **77**[2] 562-70 (1994).

HIGH TEMPERATURE BENDING STRENGTH AND FRACTURE ENERGIES OF THE TAPE-CAST SILICON NITRIDE WITH β-Si₃N₄ SEED ADDITION

Yu-Ping Zeng, Naoki Kondo, Guo-Jun Zhang, Hideki Kita and Tatsuki Ohji

Synergy Materials Research Center, National Institute of Advanced Industrial Science and Technology (AIST), Nagoya 463-8687, Japan

High temperature bending strength and fracture energies of a tape-cast silicon nitride with and without rod-like β-Si₃N₄ seed addition were measured. The experiment results showed that Lu_2O_3-SiO_2 additives were useful to improve the high temperature bending strength. This was attributable to the formation of a higher melting point grain boundary phase, which was extensively crystallized during sintering. Since the seeded and tape-cast silicon nitride has an anisotropic microstructure, where the elongated grains grown from seeds were preferentially oriented parallel to the casting direction. The strength and fracture energies of the Si₃N₄ were strongly dependent on the tape casting direction. When a stress was applied along with the grain alignment, the bending strength of the samples parallel to tape casting direction measured at 1500°C was 738 MPa, which was almost the same as the room temperature bending strength 739 MPa, and the fracture energy of the seeded and tape-cast Si₃N₄ was improved from 301 J/m² at room temperature to 781 J/m² at 1500°C. The large high temperature bending strength and fracture energies were attributable primarily to the aligned fibrous Si₃N₄ grains and sintering additives.

INTRODUCTION

Silicon nitride (Si₃N₄) ceramics is a good structural material for high temperature application due to its excellent mechanical properties. However, the pure Si₃N₄ is hard to be densifed due to its low self-diffusivity of this type of covalent material. Normally, oxides, such as Y_2O_3, Al_2O_3, Yb_2O_3, ZrO_2, Nd_2O_3 and Dy_2O_3, etc.[1-2] are added to improve the sinterability. Si₃N₄ can be densified through forming a glassy grain boundary with additives during sintering. However, the most grain boundary glass phases soften at temperature >1200°C,[3] which degrades the high temperature strength of Si₃N₄. Many efforts have been made improve the high temperature bending strength of Si₃N₄. The refractory additives and grain boundary phase crystallization have been considered as one of the useful methods[4]. The improvements of the high bending strength of Si₃N₄ were achieved via controlling the sintering additives, the composition of grain boundary and sintering program.[5] At present, oxide lutetia (Lu_2O_3) is considered as the best sintering additive for improving the high temperature strength of Si_3N_4.[6] The formation of a high melting point $Lu_4Si_2O_7N_2$ grain boundary phase that can be extensively crystallized during sintering which improve the high temperature mechanical properties.

Besides the adjustment of the sintering additives and grain boundary compositions, many researchers have tried to control grain size, and grain morphology to enhance the mechanical properties of Si_3N_4 ceramics.[7] It is well known that the high fracture toughness of silicon nitride materials is a consequence of the "*in situ*" self-reinforcing mechanism associated with the fibrous β-Si_3N_4 grains developed during sintering.[8] Using rod-like β-Si_3N_4 seed, Hirao *et al.*[9] prepared Si_3N_4 ceramics with fibrous grain alignment via tape casting processing and obtained a high fracture toughness (~11 MPa·m$^{1/2}$) as well as a high fracture strength (~1.1 GPa) when a stress was applied parallel to the fibrous grain alignment. Kondo *et al.*[10] succeeded in the alignment of fibrous Si_3N_4 grains using a superplastic-forging and sinter-forged techniques, also resulting in excellent mechanical properties; for example, the fracture toughness and strength of superplastically sinter-forged Si_3N_4 were ~8 MPa·m$^{1/2}$ and 2.1 GPa, respectively. Inagaki *et al.*[11] developed a porous silicon nitride via seeding and tape casting process; the fracture energy of the aligned porous Si_3N_4 was ~500 J/m^2 at room temperature, which was 7 times larger than that of a dense Si_3N_4 with randomly oriented fibrous grains. High fracture toughness is attributable to grain alignment, which could increase the opportunity for crack deflection and bridging. It is expected that the combination of a high melting point grain boundary phase and a fibrous grain alignment would lead to high fracture energy and high strength at elevated temperatures. The purpose of our research is to investigate high temperature strength and fracture energies of the tape-cast Si_3N_4 ceramics with and without seed addition

EXPERIMENTAL PROCEDURE

The starting powders used in the experiments were Si_3N_4 (SN-E10, UBE Industries, Tokyo, Japan), Lu_2O_3 (99.9% purity, Shinetsu Chemical Co., Ltd., Tokyo, Japan), and SiO_2 (High Purity Chemical Co., Ltd., Tokyo, Japan). Rod-like β-Si_3N_4 seeds were prepared in our laboratory. Figure 1 shows the SEM micrograph of the rod-like β-Si_3N_4 seeds. The mixed powder of 3 wt% β-Si_3N_4 seed, 90 wt% Si_3N_4, 1 wt% SiO_2, and 9 wt% Lu_2O_3 was ball-milled with solvent and dispersant for 24 h, then adding binder and plasticizer and ball milling for another 48 h. The slurry was dispersed in an ultrasonic bath for 5 minutes to break up soft agglomerates and degassed in vacuum before tape casting. The tapes were prepared with a tape casting machine (Sansho Industrial, Co., Ltd., Osaka, Japan). The casting speed was 20 cm/min. The cast slurry was dried on the drying chamber, whose temperature can be controlled in three different stages of 45°C, 55°C, and 70°C. The thickness of the dried green tape was about 100 μm. The green bodies of Si_3N_4 obtained through debonding, cutting, laminating and CIPing. They were then sintered at

1950°C in 10 atm N$_2$ atmosphere for 6 h. The rectangular shape specimens with dimension 3 mm

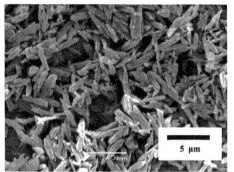

× 2 mm × 20 mm were machined to measure the bending strength at room temperature (6 specimens) and 1500°C (4 specimens). A three-point bending test (span 16 mm) was used in the experiment. The chevron- notched beam (CNB) technique was employed to measure the room temperature and 1500°C fracture energy. The rectangular shape specimens with dimension 3 mm × 4 mm × 40 mm were used in the fracture energy measurement and the displacement rate was 0.01 mm/ min. Figure 2 shows the schematic diagram of the chevron-notched beam specimen. A silicon carbide bending fixture was used in the testing system, which has high rigidity and makes crack growth stable. The microstructure was characterized by the scanning electron microscope (SEM) of specimens with polished and plasma -etched surface. X-ray diffraction (XRD) was used to determine phase composition.

Fig. 1. SEM photograph of the β-Si$_3$N$_4$ seeds

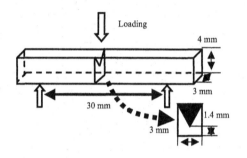

Fig. 2. Schematic diagram of the chevron–notched beam specimen.

RESULTS AND DISCUSSION

Figure 3 shows the XRD patterns of the 9 wt% Lu$_2$O$_3$-1 wt% SiO$_2$-Si$_3$N$_4$ with and without Si$_3$N$_4$ seed addition. Since the amount of the additives for the tape-cast Si$_3$N$_4$ with and without seed addition was the same, they have the same composition of the main phases Si$_3$N$_4$ and Lu$_4$Si$_2$O$_7$N$_2$. Figure 4 shows the typical SEM micrographs of the tape-cast Si$_3$N$_4$ with and without seed addition. Figure 4 (a) and (b) indicate that the seeded and tape-cast Si$_3$N$_4$ is an anisotropic material, where the fibrous Si$_3$N$_4$ grains aligned along with the tape casting direction. As previously reported[12], whiskers or particles with anisotropic shapes are easily rotated and re-arranged during tape casting; the aligned rod-like β-Si$_3$N$_4$ seeds enhance β-Si$_3$N$_4$ unidirectional grain growth. On the other hand, the SEM micrograph (Fig. 4 (c)) of the Si$_3$N$_4$ without rod-like β-

Si$_3$N$_4$ seed addition shows the isotropic microstructure. Figure 5 shows the bending strength of the tape-cast Si$_3$N$_4$ with and without rod-like seed addition at room temperature and 1500°C.

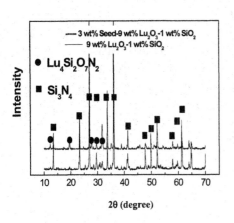

Fig. 3. XRD pattern of the tape-cast Si$_3$N$_4$

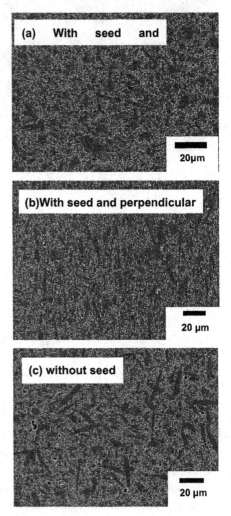

Since the seeded and tape-cast Si$_3$N$_4$ was an anisotropic material, the bending strength of the material was also strongly dependent on the applied stress direction. The results indicate that with seed addition the bending strength of the specimens parallel to the casting direction is much larger than that of the perpendicular direction. When a stress was applied parallel to the fibrous grain alignment, the bending strength of the Si$_3$N$_4$ with seed addition at 1500°C was 738 MPa, which was almost the same as the room temperature bending strength 739 MPa. However, when a stress was applied perpendicularly to the fibrous grain alignment, the bending strength of the seeded Si$_3$N$_4$ was even lower than those of the un-seeded Si$_3$N$_4$. The high bending strength of the seeded and tape-cast Si$_3$N$_4$ at 1500°C is attributable to the

Fig. 4. SEM micrographs of the tape-cast Si$_3$N$_4$ with and without rod-like seeds addition (a) with seed and parallel, (b) seed and perpendicular, and (c) without seed

formation of the high melting point $Lu_2Si_2O_7N_2$ phase and the fibrous $\beta\text{-}Si_3N_4$ grain alignment.

Crack propagation behavior is significantly dependent on the fibrous $\beta\text{-}Si_3N_4$ grain alignment. If the crack propagates perpendicularly to the grain alignment, the intragranular fracture will dominate the fracture mode that enhance crack bridge effects. The large fibrous Si_3N_4 grains increased the opportunities for crack bridging and crack deflection[13]. If the crack propagates parallel to the fibrous grain alignment, intergranular fracture is very easy to occur, resulting in the low bending strength. This is the main reason that all the seeded and tape-cast specimens exhibited high bending strength when a stress was applied along with the fibrous grain alignment. Figure 6 shows the typical load–deflection curves of the

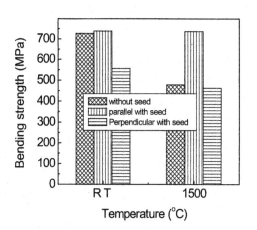

Fig. 5. Bending strength of the tape-cast Si_3N_4 at different temperatures

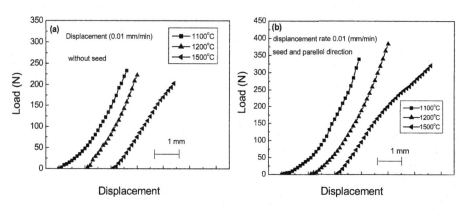

Fig. 6. Typical load –deflection curves of the tape cast Si_3N_4 at different teperatures
(a) without seed addition, and (b) seed nd parallel direction

tape cast Si_3N_4 with seed and without seed. The curves exhibit a linear deformation behavior up to facture for the tape-cast Si_3N_4 with and without seed addition at temperatures lower than 1500°C. At 1500°C, the transition of the load-deflection curve from linear to nonlinear behavior was observed. The deformation behavior was tightly related to the grain boundary phase at high temperature. Although the XRD analyses indicate formation of the high melting point grain boundary phase $Lu_4Si_2O_7N_2$, the complete crystallization of the grain boundary phase is considered to be rather difficult.[5] Softening of the remanent glassy phase at high temperatures leads to degradation of deformation resistance. Guo et al.[6] indicated the nonlinear deformation of the hot-pressure Si_3N_4 with doped Lu_2O_3 appeared at 1500°C, and the deformation behavior was dependent on the amount of the doped Lu_2O_3.

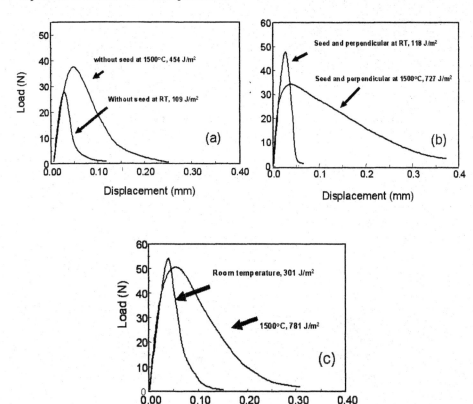

Fig. 7. Load –displacement diagrams of the chevron-notched bend (CNB) tests for the tape-cast Si_3N_4 (a) without seed, (b) seed and perpendicular direction, and (c) seed and parallel direction.

Figure 7 shows the load–displacement diagrams of the chevron-notched bend (CNB) tests for the tape-cast Si_3N_4. The diagrams of the tape-cast Si_3N_4 show a relative smooth L–D curve. Kondo et al.[14] indicated that the fracture energies of Si_3N_4 were strongly affected by the temperatures and displacement rates. In this research, we selected 0.01 mm/min as the displacement rate in all measurements. The fracture energy, γ_{eff}, is defined as follows:

$$\gamma_{eff} = \frac{W_{WOF}}{2A}$$

where W_{WOF} is the energy under the L–D curve and A is the area of the specimen web portion. Three specimens were used and stable crack growth until the completion of the test was obtained. It is observed that the area under the L–D is strongly dependent on the measuring temperature, and the area under L–D curve at 1500°C is larger than that of the room temperature. The results indicate that the fracture energies of the Si_3N_4 increase with rod-like Si_3N_4 seed addition. Furthermore, the fracture energies of the seeded Si_3N_4 are also dependent on the applied stress direction; the fracture energies of the seeded and tape-cast Si_3N_4 with an applied stress parallel to the fibrous grain alignment are always larger than those with an applied stress perpendicular to the fibrous grain alignment. When a stress was applied parallel to the fibrous grain alignment, the fracture energy of the seeded Si_3N_4 at room temperature was 301 J/m^2, which was almost 3 times larger than that of the un-seeded Si_3N_4. The fracture energy at 1500°C was 781 J/m^2, which was also larger than that of the un-seeded Si_3N_4 of 454 J/m^2.

CONCLUSIONS

Bending strength and fracture energies of the tape-cast Si_3N_4 with and without rod-like β-Si_3N_4 seed addition were investigated at room temperature and 1500°C. The results showed that with the seed addition the bending strength and fracture energies of the tape cast Si_3N_4 were greatly improved. When a stress was applied parallel to the fibrous grain alignment, the bending strength of the seeded and tape-cast Si_3N_4 at 1500°C was 738 MPa, which was almost the same as the room temperature bending strength 739 MPa. The aligned fibrous Si_3N_4 grains and the interlocking microstructure of the seeded Si_3N_4 cause these the high fracture energies.

REFERENCES

[1]I.C. Huseby and G. Petzow, "Influence of Various Densifying Additives on Hot-Pressed Si_3N_4", Powder Metallurgy International, 6, 17–19 (1974)
[2]W.A. Sanders and D.M. Miekowski, "Strength and Microstructure of Si_3N_4 Sintered with Rare-Earth Oxide Additions," American Ceramic Society Bulletin, 64 [2] 304–309 (1985).
[3]F.F. Lange, "High Temperature Strength Behavior of Hot-Pressed Si_3N_4: Evidence for Subcritical Crack Growth," Journal of the American Ceramic Society, 57 [2] 84–87 (1974)

[4]A. Tsuge, K. Nishida, and M. Komatsu, "Effect of Crystallizing the Grain-Boundary Glass Phase on the High-Temperature Strength of Hot-Pressed Si_3N_4 Containing Y_2O_3," *Journal of the American Ceramic Society*, 58 [7–8] 323–26 (1975).

[5]T. Nishimura and M. Mitomo "High Temperature Strength of Silicon Nitride Ceramics with Ytterbium Silicon Oxynitride," *Journal Materials Research*, 12 [1] 203-209 (1997).

[6]S. Guo, N. Hirosaki, Y. Yamamoto, T. Nishimura, and M. Mitomo, "Improvement of High-temperature Strength of Hot-pressed Sintering Silicon Nitride with Lu_2O_3 Addition", *Scripta Materialia*, 45, 867-874 (2001).

[7]J.S. Wallace and J.F. Kelly, "Grain Growth in Si_3N_4," *Key Engineering Materials*, 89–91, 501–505 (1994).

[8]P.F. Becher, E.Y. Sun, K.P. Pluckner, K. B. Alexander, C.-H. Hsueh, H.-T. Lin, S.B. Waters, C. G. Westmoreland, E.-S. Kang, K. Hirao, and E. Brito, "Microstructural Design of Silicon Nitride with Improved Fracture Toughness: I, Effects of Grain Shape and Size," *Journal of the American Ceramic Society*, 81 [11] 2821–30 (1998).

[9]K. Hirao, T. Nagaoka, M.E. Brito and S. Kanzaki, "Microstructure Control of Silicon Nitride by seeding with rod-like beta-Silicon Nitride particles," *Journal of the American Ceramic Society*, 77 [7] 1857-62 (1994)

[10]N. Kondo, Y. Suzuki, and T. Ohji, "Superplastic Sinter-Forging of Silicon Nitride with Anisotropic Microstructure Formation," *Journal of the American Ceramic Society*, 82 [4] 1067 – 69 (1999).

[11]K. Hirao, M. Ohashi, M.E. Brito, and S. Kanzaki, "Processing Strategy for Producing Highly Anisotropic Silicon Nitride", *Journal of the American Ceramic Society*, 78 [6] 1687–90 (1995).

[12]H. Imamura, H. Kiyoshi, E.B. Manuel, T. Motohiro, and K. Shuz, "Further Improvement in Mechanical Properties of Highly Anisotropic Silicon Nitride Ceramics", *Journal of the American Ceramic Society*, 83 [3] 495–500 (2000).

[13]T. Ohji, K. Hirao, and S. Kanzaki, "Fracture Resistance Behavior of Highly Anisotropic Silicon Nitride", *Journal of the American Ceramic Society*, 78 [11] 3125–28 (1995).

[14] N. Kondo, M. Asayama, Y. Suzuki, and T. Ohji, "High-Temperature Strength of Sinter-Forged Silicon Nitride with Lutetia Additive," *Journal of the American Ceramic Society*, 86 [8] 1430–1 432 (2003).

SUPERPLASTICITY OF THE NANOSTRUCTURED BINARY SYSTEMS OF ZIRCONIA-ALUMINA-SPINEL CERAMICS BY SPARK PLASMA SINTERING PROCESS

Xinzhang Zhou, Dustin M. Hulbert, Joshua D. Kuntz, Javier E. Garay and Amiya K. Mukherjee

Department of Chemical Engineering and Materials Science
University of California, Davis, 95616

ABSTRACT

Three biphase composites (zirconia-alumina, alumina-spinel and zirconia-spinel) of the zirconia-alumina-spinel ceramic were processed separately from nanosized alumina, zirconia and magnesia powders. The samples became nearly fully dense after spark plasma sintering at 1050-1100 °C, with the grain sizes between 50 nm and a few hundreds of nanometers. Specimens cut from the sintered compacts were tested under compression at elevated temperatures (1300-1500 °C) and at strain rates ranging from 10^{-1} to 10^{-5} s^{-1}. Strain sensitivities and activation energies of the biphase composites were compared with that of the triphase composite. Additionally, microstructures of the binary systems before and after the superplastic deformation were studied by SEM. Lastly, grain boundaries and phase boundaries that were most prone to sliding were investigated in conjunction with mechanical parameters in order to shed more light on the rate controlling mechanisms.

INTRODUCTION

Superplasticity was first widely studied in metals and alloys. The constitutive relation for superplastic deformation (SPD) usually takes the form of Mukherjee-Bird-Dorn Equation[1]:

$$\dot{\varepsilon} = A\frac{DGb}{kT}(\frac{b}{d})^{p}(\frac{\sigma}{G})^{n}e^{-\frac{Q}{RT}} \tag{1}$$

in which G is the elastic shear modulus, b is the Burger's vector, k is the Boltzmann's constant, T is absolute temperature, d is grain size, p is the grain-size dependence coefficient, n is the stress exponent, Q is activation energy, D is the diffusion coefficient and R is the gas constant. The inverse of n is the strain rate sensitivity, m. Grain boundary sliding is generally the predominant mode of deformation during the superplastic flow. Plastic deformation by grain-boundary sliding is generally characterized by $n=2$ ($m=0.5$) and an activation energy that is either equal to the activation for lattice diffusion or to the activation energy for grain-boundary diffusion.

From Equation (1), it is clear that for elevated temperature deformation at a constant temperature, high strain rate is more easily realized in specimens with smaller grains. With the development of ceramic processing, the particle sizes are now made smaller and smaller into the nanometer range and therefore so are the grain sizes in dense compacts. Superplasticity in ceramics has also been studied since the first observation of the other fine-structure superplasticity in yttria-stabilized tetragonal zirconia (YTZP) by Wakai in 1986[2]. A number of fine-grained polycrystalline ceramics have demonstrated superplasticity, such as YTZP[3], magnesia-doped alumina[4], and alumina reinforced YTZP[5]. Unfortunately, the superplastic temperatures were typically above 1450°C and the strain rates were relatively low (10^{-4} s^{-1} or lower). Recently, Kim, et al[6] realized a high strain rate of 0.1 s^{-1} in zirconia-alumina-spinel triphase composite (volume ratio 4:3:3), but at a rather high temperature of 1650°C.

More recently, superplasticity of the triphase ceramic composite at temperatures as low as 1350 °C was demonstrated in samples processed by spark plasma sintering (SPS)[7]. Because of the rather low sintering temperatures (1100-1200°C) and the very short sintering time (a few minutes), the grain sizes were about 50 nm to 100 nm. However, the complicated grain boundaries and phase interfaces in the triphase composite make it difficult to study the rate-controlling mechanisms during the high temperature deformation. In the triphase ceramic composite, there are three grain boundaries (zirconia-zirconia, alumina-alumina, and spinel-spinel), and three interphases (zirconia-alumina, zirconia-spinel and spinel-alumina) in the triphase ceramic. In this study, three biphase composites (zirconia-alumina, zirconia-spinel and alumina-spinel) of the same volume ratio as in the triphase ceramic were processed by SPS and later tested at various elevated temperatures and strain rates. This aims to deconvolute the rate-controlling mechanisms in the triphase ceramic composite.

MATERIALS PROCESSING AND EXPERIMENTS

(1) Raw Materials
   The starting nanocrystalline powders are listed in Table I. Starting from the nanocrystalline powders SPS has made ceramics with nearly full density, grains of nanometer sizes and uniform distribution of different phases[7, 8].

Table I Raw Materials and Their Properties

Powder	Provider	Grain size (nm)	Phase
Zirconia	TZ3Y, Tosoh (Tokyo, Japan)	25	Tetragonal
Alumina	Nanotechnologies, (Austin, TX)	15	γ
Magnesia	NEI, (Piscataway, NJ)	40	Cubic

(2) Powder Processing
   γ alumina powder was first high energy ball milled (HEBM) for 24 hours in a Spex 8000 mixer mill in a tungsten carbide (WC) vial. It was reported[8] that HEBM builds up intensive strain in the nanopowder and helps sintering and phase transformation from γ to α during

sintering. One wt% polyvinyl alcohol (PVA) was added as a dry milling agent to prevent severe agglomeration. A heat treatment in the air at 350°C for 3 hours after HEBM removed the PVA from the powder. Combinations of the three powders (zirconia-33wt% alumina, zirconia-22wt% alumina-9wt% magnesia, and alumina-13wt% magnesia) were mixed in methanol and ball milled for 24 hours in zirconia media. The slurries were dried in air and heated at 350°C for 3 hours to remove any residual organics. After sieving to remove the particles over 100 μm, the powders were unidirectionally pressed at 30 MPa in a graphite die. The relative green densities were around 45%-50%.

(3) SPS

SPS is a new sintering method with moderate pressures and high temperatures[9]. Spark plasma is generated between the particles by electrical discharge at the onset of on-off DC pulsing. The on-off DC pulses may result in spark plasma, spark impact pressure, Joule heating and electrical field diffusion. SPS can rapidly consolidate the green compact to nearly theoretical density. In SPS, rapid heating rate (normally a few hundreds °C/min), moderate pressure and short sintering time (minutes) makes it possible to consolidate fully dense ceramic with retarded grain growth.

In this study, Dr. Sinter 1050 (Sumitomo Coal Mining Co., Japan) was used to sinter the green compacts to full density. The heating rate was 300°C/min from the room temperature to 600°C and then in 2 minutes the samples were ramped to the desired temperatures. A unidirectional pressure of 64 MPa was applied during the sintering. After holding at the desired temperatures for 3 minutes, the power was shut down and the specimens were cooled in vacuum. An optical pyrometer focusing on the graphite die monitors the temperatures. The whole process was controlled by a PC interface LabView program. The final densities of the sintered compacts were measured by the Archimedes method with deionized water as the immersion liquid. The theoretical densities were calculated by the rule of mixtures.

(4) Mechanical Testing and Characterization

The sintered samples were cut and then polished to a bar shape about 3 by 3 by 5 millimeters. Mechanical tests were carried out on the PC Labview controlled MTS-810 loading frame. Once the samples were heated up in the air to the desired testing temperatures, strain rate-jump tests were conducted on each specimen at a certain constant temperature. The specimen was deformed at a constant strain rate until a steady load is registered before going to the next strain rate. In our tests, each specimen was deformed to a strain of 0.1 in each step to make sure that a steady flow stress was obtained. The actual load and displacement were recorded and changed into stress and strain values for further analysis.

Phases were identified by X-ray diffraction (XRD) using K_α radiation. An FEI XL30-SFEG high resolution scanning electron microscope (SEM) was used to observe the microstructures.

RESULTS AND DISCUSSIONS

(1) SPS

In order to obtain the smallest grain size and nearly full density, different SPS temperatures and dwelling times were used. In Table II, the optimal SPS parameters and densities of the specimens are listed. The sintering temperatures used were minimized while still obtaining near fully dense specimens. All relative densities were over 98% and the grain sizes were estimated from SEM images of the corresponding fracture surfaces.

Table II SPS Parameters and the Densities

Materials	SPS Temperature (°C)	Sintering Time (minutes)	Density (g/cm³)	Theoretical Density (%)	Grain Size (nm)
Alumina-zirconia	1100	3	5.07	98.0	(70, 50)
Alumina-spinel	1050	3	3.73	98.7	(100,400)
Zirconia-spinel	1050	3	4.91	98.1	(50, 400)

(2) XRD Results

From Fig. 1, it is clear that during SPS, γ alumina changed into α alumina in zirconia-alumina, into α alumina and spinel in alumina spinel and into spinel in zirconia-spinel and neither MgO nor γ alumina phase were detected by XRD.

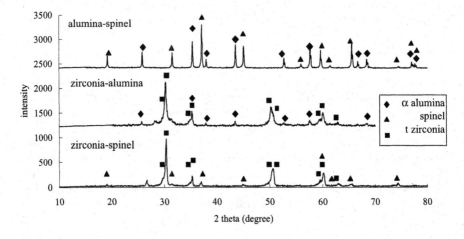

Fig. 1 XRD of the specimens after sintering
(t zirconia means tetragonal zirconia)

(3) Superplasticity and the Microstructures

According to Equation (1), the slope of a log (flow stress) vs. log (strain rate) curve can give the strain rate sensitivity m. Fig. 2 shows the superplastic behavior for zirconia-alumina composite. The slope of each line is around $m=0.5$, i.e. a stress exponent of the strain rate equals to 2. It clearly shows superplasticity in the temperature and strain rate range with the stress exponent n around 2. A similar graph (Fig. 3) can be drawn for zirconia-spinel. SEM images of the deformed specimen (Fig. 4) show that the equiaxed grains are round at the corners after deformation although the starting grains have sharp corners. Therefore the grain or interface boundaries experienced substantial sliding during SPD. Furthermore, the samples did not fracture during the strain-rate jump tests up to the total (0.51) strain level investigated.

However, alumina-spinel did not show superplasticity in temperatures ranging from 1300°C to 1490°C, and strain rates varying from 10^{-5} to 10^{-1} s^{-1}. The specimens all fractured at true strain of about 0.2 to 0.3, even at strain rates as low as 10^{-5} s^{-1} and at temperatures as high as 1490°C and as low as 1350°C. One of the possible reasons is that the grains are too big for the superplasticity to manifest itself. Another possible reason might be related with the interface structure of alumina-spinel. As shown in Fig. 4, it is clear that the spinel grains grew fastest during the high temperature deformation. After the testing at 1450°C, grain size is over 1 micron for the spinel compared to the submicrons grains of zirconia and alumina.

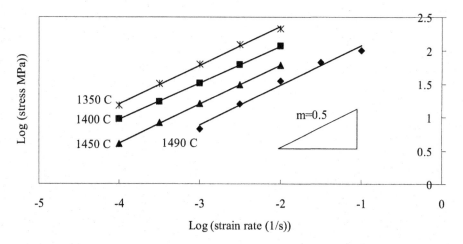

Fig. 2 The stress-strain rate curves used to determine strain rate sensitivity m for zirconia-alumina

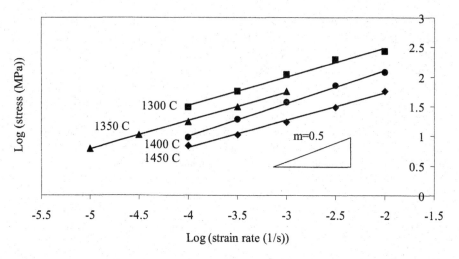

Fig. 3 The stress-strain rate curves used to determine strain rate sensitivity *m* for zirconia-spinel

The grain sizes and shapes of the biphase and triphase composites before and after the SPD are shown in Fig. 4. In the as-sintered specimens, grains are smallest in the triphase ceramic and in zirconia-alumina binary ceramic (from 25 nm to around 40-50 nm). Grain growth of alumina and spinel are substantial in the sintering process (from 15 nm and 40 nm to around a few hundred nanometers, respectively). During the mechanical testing at 1450°C, the grain growth varied for the three phases. Zirconia grew slowest (to 200-300 nm), while alumina grew quickly (up to 1μm) and spinel grew fastest (up to over 5 μm), especially in the alumina-spinel system.

(A) Zirconia-alumina – as sintered

(B) Zirconia-alumina - after SPD

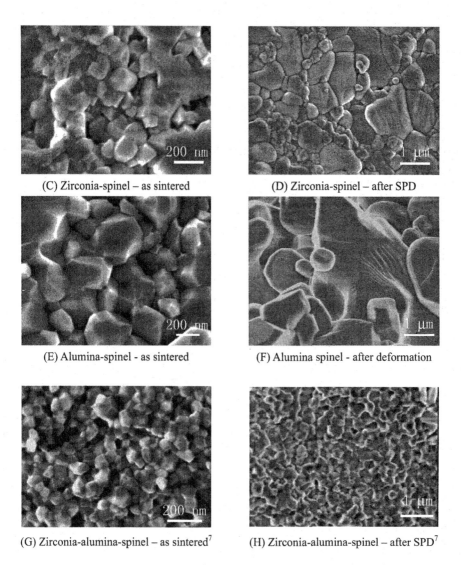

(C) Zirconia-spinel – as sintered

(D) Zirconia-spinel – after SPD

(E) Alumina-spinel - as sintered

(F) Alumina spinel - after deformation

(G) Zirconia-alumina-spinel – as sintered[7]

(H) Zirconia-alumina-spinel – after SPD[7]

Fig. 4 Facture surfaces of the specimens before and after SPD

(5) Activation Energy Q

From Equation (1), the activation energy Q can be derived from the slope of the Ln (strain rate) vs. $1/T$. At constant stress the activation energy of zirconia-alumina is calculated from Fig. 5. A similar graph can be drawn for zirconia-spinel. It was found that the Qs for these two composites are 597 kJ/mol and 522 kJ/mol, respectively.

The activation energies Q of the binary and triphase composites are shown in Fig. 6. The activation energy of zirconia-spinel and zirconia-alumina are less than that of the triphase composite. The activation energy of alumina-spinel is not shown because at the temperatures and strain rate tested, it did not demonstrate superplasticity. So far it can be proposed that the superplasticity of the triphase system is directly related with grain-boundary sliding of zirconia-alumina and zirconia-spinel, with alumina-spinel acting to hinder the grain growth of the other two phases.

It is interesting to notice that the activation energy of zirconia-spinel is lower than that of zirconia-alumina, although the spinel grain size is larger than that of the alumina grains while the zirconia grains are of similar sizes in both cases. The interface microstructures might play an important role. Further TEM study may shed some light on this question.

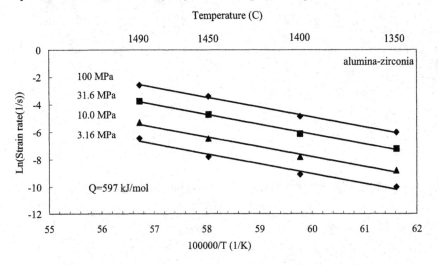

Fig. 5 Ln (Strain rate) vs. temperature relation to determine the activation energy of zirconia-alumina

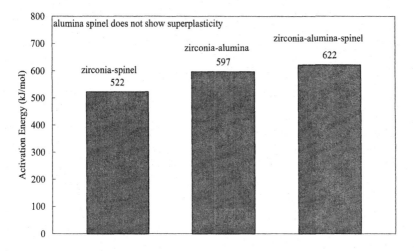

Fig. 6 Activation energy of the binary and triphase ceramic composites
(Data for the triphase is from ref. 7)

CONCLUSIONS

(1) Three biphase ceramic composites (zirconia-alumina, alumina-spinel and zirconia-spinel) with nearly full density and grain sizes ranging from submicron to below 40 nm, were processed by SPS.

(2) Zirconia-alumina and zirconia-spinel exhibited superplasticity in compressive tests at temperatures between 1300°C and 1500°C in the strain rates between 10^{-5} to 10^{-1} s^{-1}, while alumina-spinel did not show superplasticity in the same testing condition. The activation energies are 522 kJ/mol and 597 kJ/mol for zirconia-spinel and zirconia-alumina, respectively.

(3) While zirconia-alumina and zirconia-spinel contribute directly more to the grain boundary sliding, alumina-spinel appears to play an important role in hindering the grain growth, contributing indirectly to the enhanced superplasticity of the triphase composite.

ACKNOWLEDGEMENT

This research is supported by the U.S. Office of Naval Research (Grant No. N00014-03-1-0148). X. Z. and D. H. are grateful to Drs. Guodong Zhan and Renguan Duan for their technical aid and helpful discussion.

REFERENCES

[1]A. K. Mukherjee, J. E. Bird and J. E. Dorn, "Experimental Correlations for High-Temperature Creep", *ASM Trans.* **62**(1), 155-79 (1969);

[2]F. Wakai, S. Sakaguchi, and Y. Matsuno, "Superplasticity of Yttria-Stabilized Tetragonal ZrO_2 Polycrystals", *Adv. Ceram. Mater.* **1**, 259-63 (1986);

[3]T. G. Nieh and J. Wadsworth, "Superplasticity Behavior of A Fine-Grained Yttria-Stabilized Tetragonal Zirconia Polycrystal (Y-TZP)", *Acta Metall. Mater.* **38**, 1121-33 (1990);

[4]Y. Yoshizawa and T. Sakuma, "Improvement of Tensile Ductility in High-Purity Alumina Due to Magnesia Addition", *Acta Metall. Mater.* **40**, 2943-50 (1992);

[5]F. Wakai and H. Kato, "Superplasticity of TZP/Al2O3 Composite", *Adv. Ceram. Mater.*, **3** 71-6 (1988);

[6]B. N. Kim, K. Hiraga, K. Morita and Y. Sakka, "A High-Strain-Rate Superplasticity Ceramic", **413**, 288-291 (2000);

[7]Joshua D. Kuntz, J. Wan and A. K. Mukherjee, "High Strain-Rate Superplastic Deformation of an $Al_2O_3/ZrO_2/MgAl_2O_4$ Nanocomposite", submitted, 2004;

[8]G. D. Zhan, J. Kuntz, J. Wan, J. Garay and A. K. Mukherjee, "A Novel Processing Route to Develop A Dense Nanocrystalline Alumina Matrix (<100 nm) Nanocomposites Material", *J. Am. Ceram. Soc.* **86**[1], 200-2 (2002);

[9]M. Omori, "Sintering, Consolidation, Reaction and Crystal Growth by the Spark Plasma System (SPS)", *Mater. Sci. Eng.* A **287**, 183-8 (2000);

INITIATION OF MATRIX CRACKING IN WOVEN CERAMIC MATRIX COMPOSITES

Madhwapati Prabhakar Rao, Michael Pantiuk and Panos G. Charalambides
Department of Mechanical Engineering, University of Maryland Baltimore County
1000, Hilltop Circle
Baltimore
MD 21250

ABSTRACT
    In this study, we employ detailed 3D finite element models of complex plain and satin weave ceramic matrix composites to address the problem of initiation of matrix cracking. Fundamental 3D elasticity boundary value problems addressing the response of these materials under the combined effects of remote biaxial tension and pure shear loads coupled with residual thermal stresses are utilized as needed to characterize the matrix micro-stresses in the vicinity of large-scale macroscopic voids. The combined stress fields are then used to determine the effects of remotely applied in-plane loads and thermal temperature gradients on the initiation of matrix cracking in the region of high stress concentrations. Selective comparison of model results with available experimental data are discussed. Extensive parametric studies have been carried out and robust failure loci predictive of the "first" matrix cracking events have been developed. The availability of such failure envelopes for woven CMCs offers for the first time a design tool for the use of such advanced systems in engineering applications.

INTRODUCTION

    Woven fabric composites can be engineered to provide high strength, stiffness and toughness in comparison with conventional monolithic materials for a wide variety of land, sea and space based applications. However these materials are characterized by a complex three-dimensional geometry and intricate microstructures [1]. As such robust solution techniques capable of predicting the micro-stress elastic fields are needed in the quantification of life limiting failures in these systems.

    When woven CMCs are loaded in-plane, micro-failure events such as matrix microcracking, fiber debonding, fiber bridging, frictional fiber pull-out and fiber tow delamination have been shown to occur [2, 3]. Overall woven CMCs are known to exhibit a nonlinear stress-strain curve characterized by a graceful ultimate failure [3]. Under the influence of high temperature gradients, the apparent strength of woven CMCs has been shown to decrease mainly due to residual stresses induced by thermal expansion mismatch of the fiber and matrix materials [4]. Most applications with these material systems are likely to be designed using the "first knee" of the general nonlinear stress-strain curve as the stress design limit. This "first knee" has been shown to coincide with the "first matrix cracking" stress identified as the proportional limit of the woven composite material [5].

    From the standpoint of engineering design, the proportional limit stress of woven CMCs is a parameter of critical importance. Experimentally established tensile stress-strain curves for a SiC/SiC plain weave system obtained by Williams International, Inc. and CCI, Inc. are shown in Figure 1 which clearly indicate that the proportional limit stress of woven CMCs indeed decreases with increasing temperature. Therefore, the objective of this work is to develop broad failure loci for a plain weave C/SiC system subjected to a combination of in-plane mechanical loads and thermal temperature gradients and in which the fibers and matrix are thermally different.

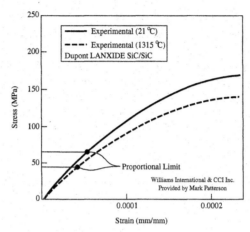

Figure 1: Experimentally established stress-strain curves for a DuPont Lanxide plain weave SiC/SiC composite. Data reported by Patterson [5, 6].

FINITE ELEMENT MODELING

In this work we model the mechanical response of complex woven CMCs employing the finite element method. Detailed three-dimensional (3D) finite element models of the symmetric unit-cells of woven CMCs are developed using 3D 8-noded isoparametric brick elements, consistent with studies reported elsewhere [7, 8, 9]. The fiber tows are modeled as transversely isotropic material entities with directional effective properties while the matrix material is modeled as an isotropic medium with dispersed porosity. Symmetry boundary conditions are enforced on the woven unit-cells as required to simulate in-plane mechanical loads and thermal expansion effects.

Geometric Modeling

An improved binary sub-cell method was developed to model the geometry of Four Harness (4HS), Five Harness (5HS) and Eight Harness (8HS) satin weave composites on the lines of the methodology suggested by Hewitt et. al. [10] without sacrificing the details about the cross-sectional shapes and undulating profiles of the fiber tows. However, this study represents a marked improvement over the previous models since discrete large matrix voids as well as dispersed matrix porosity are incorporated into hierarchical models capable of predicting the macroscopic response while maintaining the integrity of the microstructure. These hierarchical models can now be used to predict micro-stress concentration sites and induced damage associated with the non-linear branch of the stress-strain curves [5, 11]. The modeling of the Plain Weave geometry follows the techniques described in [7]. Figure 2 shows a library of the finite element models of the symmetric unit-cells of the woven composites modeled as part of an ongoing research effort.

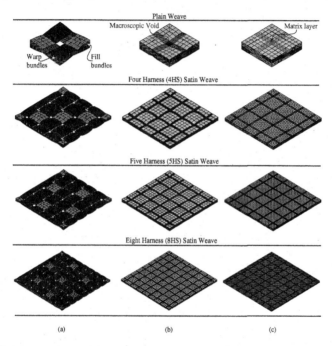

Figure 2: Geometric finite element models of the symmetric unit-cells of various woven composites. (a) Fiber bundle architecture of plain and satin weave composites. (b) Ceramic matrix layer deposited on woven fibers via the Chemical Vapor Infiltration (CVI) technique. (c) Polymer matrix layer deposited on woven fibers.

The analysis presented as part of this work was performed on a plain weave C/SiC system with non-dimensional spatial variables listed in Table I . The half-period of fiber tow undulation a was chosen as the characteristic spatial dimension to normalize the spatial variables. This approach is consistent with the non-dimensionalization procedure developed in [7, 8]. A representative finite element mesh made with the geometric input parameters of

Table I is shown in Figure 3. The actual mesh used in the parametric studies was discretized with 17920 3D 8-noded isoparamteric brick elements. The model consisted of 20611 finite element nodes resulting in a total of 61833 degrees of freedom. The physical properties of the plain weave C/SiC system modeled in this work are given in Table II.

Table I: Microstructural Geometry Parameters.

Parameter	Description	Value
$\hat{a}$	a/a	1.0
$\hat{b}$	b/a	0.10
$\hat{g}$	g/a	0.15
$\hat{t}$	t/a	0.04
$\hat{h}$	h/a	$2\hat{b}$

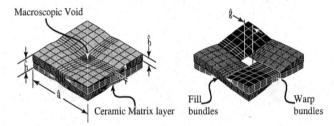

Figure 3: Finite element mesh of the plain weave C/SiC system studied in this work.

Material Properties

The microstructure of woven CMCs is hierarchical as depicted in the micrograph of Figure 4. The finite element models designed as part of this study account for such heterogeneity in the microstructure employing the robust four-step homogenization scheme developed by Kuhn and Charalambides [8]. A graphical description of the four-step homogenization procedure is shown in Figure 5. The homogenization scheme is applied to obtain the effective properties of the fiber tow which are then used in conjunction with proper material property rotation as input to the global finite element model for characterizing the effective macroscopic response of woven composites.

In Step # 1 of the homogenization scheme (see Figure 5 (b)), the fiber and and fiber coating are homogenized into an effective orthotropic fiber employing the Composite Cylinders Assemblage model. During Step # 2 of this homogenization scheme (see Figure 5 (c)), the bundle matrix and bundle porosity are combined into an effective isotropic bundle matrix by employing the porosity micromechanics model. Step # 3 of the homogenization scheme,

Table II: Physical properties of the plain weave C/SiC system studied. These properties were provided by Ceramic Composites, Inc.

Entity	Description
Fiber	Toray T-300 1K tow carbon fiber
Matrix	SiC
Tensile Strength	$\simeq$ 300 MPa
Fiber Content	40% by volume
Process	Chemical Vapor Infiltration (CVI)
Modulus	Not Available
Composite Proportional Limit	80 MPa (Tensile)

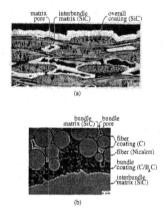

Figure 4: Micrographs of a SiC/SiC plain weave composite. (a) Sectional view of the laminate clearly showing dispersed matrix porosity. (b) Close-up view of a single fiber bundle cross-section illustrating a hierarchical microstructure.

once again employs the CCA model to homogenize the effective fibers and effective matrix into an orthotropic fiber tow (see Figure 5 (d)). The final step (see Figure 5 (e)) of this homogenization scheme, converts the effective orthotropic fiber and the bundle coating into an orthotropic material entity (fiber tow) with coating using the CCA model. The matrix between the fiber tows (inter-bundle matrix) and the associated porosity are homogenized into an effective isotropic matrix using the porosity micromechanics model.

The material micro-constituents viz., the fibers, fiber coating, bundle matrix, bundle coating, bundle porosity, inter-bundle matrix and inter-bundle matrix porosity are represented through the volume fractions C_f, C_{fc}, C_{bm}, C_{bc}, C_{bp}, C_m and C_{mp} respectively and

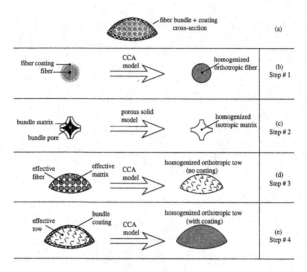

Figure 5: The four-step homogenization procedure developed by Kuhn and Charalambides [8].

are subject to the following consistency conditions:

$$C_f + C_{fc} + C_{bm} + C_{bc} + C_{bp} = 1$$
$$C_m + C_{mp} = 1 \tag{1}$$

Table III: Microstructural Material Parameters.

	Elastic Modulus	Poisson's Ratio	Shear Modulus	CTE	Volume Fraction
Fiber	$\hat{E}_f = 0.8$	$\hat{\nu}_f = 0.43$	$\hat{G}_f = 0.28$	$\hat{\alpha}_f = 0.3$	$C_f = 0.75$
Fiber Coating	$\hat{E}_{fc} = 0.125$	$\hat{\nu}_{fc} = 0.3$	$\hat{G}_{fc} = 0.05$	$\hat{\alpha}_{fc} = 1.0$	$C_{fc} = 0.05$
Bundle Matrix	$\hat{E}_{bm} = 1.0$	$\hat{\nu}_{bm} = 0.3$	$\hat{G}_{bm} = 0.385$	$\hat{\alpha}_{bm} = 1.0$	$C_{bm} = 0.14$
Bundle Coating	$\hat{E}_{bc} = 0.25$	$\hat{\nu}_{bc} = 0.25$	$\hat{G}_{bc} = 0.1$	$\hat{\alpha}_{bc} = 1.0$	$C_{bc} = 0.05$
Bundle Porosity	–	–	–	–	$C_{bp} = 0.01$
Inter-Bundle Matrix	$\hat{E}_m = 1.0$	$\hat{\nu}_m = 0.3$	$\hat{G}_m = 0.385$	$\hat{\alpha}_m = 1.0$	$C_m = 0.9$
Matrix Porosity	–	–	–	–	$C_{mp} = 0.1$

All calculations are performed in a non-dimensional environment consistent with [8, 12]. The inter-bundle matrix is chosen as the characteristic material. Table III describes the normalized material property data input to the finite element model. In Table III the "hat" notation $(\hat{\ })$ on a modulus signifies that the property has been normalized with respect to the matrix elastic modulus E_m, whereas the "hat" notation $(\hat{\ })$ on a Coefficient of Thermal Expansion (CTE) value implies that it has been normalized with respect to the matrix CTE α_m. The material micro-constituent input data presented in Table III is employed to homogenize the fiber tow and inter-bundle matrix material. The effective properties of the fiber tows and matrix material computed using the homogenization scheme developed by Kuhn and Charalambides [8] are shown in Table IV.

Table IV: Meso-scopic Output.

Effective Tow	Effective Matrix
$\hat{E}_{11} = 0.7591$	$\hat{E}_{\overline{m}} = 0.810$
$\hat{E}_{22}^l = 0.636$	$\hat{G}_{\overline{m}} = 0.314$
$\hat{G}_{12} = 0.239$	$\hat{\nu}_{\overline{m}} = 0.288$
$\hat{G}_{23} = 0.239$	$\hat{\alpha}_{\overline{m}} = 1.0$
$\hat{\nu}_{12} = 0.397$	
$\hat{\nu}_{23}^l = 0.379$	
$\hat{\alpha}_{11} = 0.4395$	
$\hat{\alpha}_{22} = 0.4263$	

The effective tow and matrix properties presented in Table IV are used to compute the transformed reduced stiffness matrix $[\overline{Q}]$ and input as the Material Property Matrix $[D]$ in the finite element model [11].

Boundary Conditions

The symmetry boundary conditions which are imposed on the woven unit-cells to simulate pure remote tension and shear loading are derived from the work of Kuhn and Charalambides [8, 12]. However, the boundary conditions for free thermal expansion are different as explained in Figure 6. As shown in Figure 6 (b), to simulate free thermal expansion, symmetry boundary conditions are imposed on the lateral faces at $\hat{x} = -\hat{a}/2$ and $\hat{y} = -\hat{a}/2$. In particular on the face at $\hat{x} = -\hat{a}/2$, the displacement of all the nodes along the global $\hat{x}$-direction is set equal to zero, i.e. $\hat{u}_x = 0$, whereas on the face of the woven unit-cell corresponding to $\hat{y} = -\hat{a}/2$, the displacement of all nodes in the global $\hat{y}$-direction is set equal to zero, i.e. $\hat{u}_y = 0$. However, on the opposite faces corresponding to $\hat{x} = \hat{a}/2$ and $\hat{y} = \hat{a}/2$, we impose constraint conditions thereby ensuring the planarity of these faces before and after deformation (as dictated by symmetry). More specifically, we require that all the nodes on the face of the woven unit-cell corresponding to $\hat{x} = \hat{a}/2$ should move by the same amount along the global $\hat{x}$-direction, i.e. $\hat{u}_x = C$, whereas all the nodes on face of the woven unit-cell at $\hat{y} = \hat{a}/2$ are forced to move by an equal amount along the global $\hat{y}$-direction, i.e. $\hat{u}_y = C$.

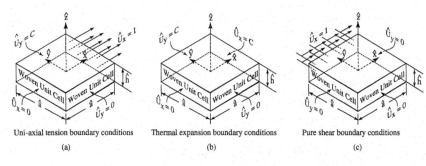

Uni-axial tension boundary conditions	Thermal expansion boundary conditions	Pure shear boundary conditions
(a)	(b)	(c)

Figure 6: Schematic representation of the boundary conditions imposed on the woven unit-cells to simulate different loading conditions. Rigid body motions in the z-direction are constrained by enforcing zero displacements in the same direction at least at one node in the FE model.

MODEL FORMULATION

In-plane mechanical loads can either be represented through the stresses σ_{xx}^{∞}, σ_{yy}^{∞} and τ_{xy}^{∞} or via the spherical loading parameters, ω, ϕ and S. Figure 7 defines these parameters and also graphically illustrates the stress fields resulting from the four fundamental boundary value problems solved in order to develop failure loci aimed at assessing the strength of woven CMCs under combined mechanical and thermal loads.

The remotely applied in-plane mechanical loads can be related to the spherical loading parameters as:

$$\begin{aligned}
\sigma_{xx}^{\infty} &= S\cos\phi\cos\omega \\
\sigma_{yy}^{\infty} &= S\cos\phi\sin\omega \\
\tau_{xy}^{\infty} &= S\sin\phi
\end{aligned} \tag{2}$$

It has been shown in previous studies [13, 14] that the total stress induced in the woven CMCs scales linearly with the remotely applied loads. Based on the principle of linear superposition, the total stress is given by:

$$\sigma_{ij}^{total} = \sigma_{ij}^{x} + \sigma_{ij}^{y} + \tau_{ij}^{xy} + \sigma_{ij}^{th} \tag{3}$$

where the ij-component of each individual stress is induced by the corresponding remotely applied mechanical or thermal load. For example, σ_{ij}^{x} is the ij-component of stress induced by σ_{xx}^{∞}. It can be shown that the total stress is related to the spherical loading parameters and induced non-dimensional stresses via a functional relationship of the form [5]:

$$\sigma_{ij}^{total} = S\hat{\Sigma}_{ij}^{total}\left(\omega, \phi, \beta, \hat{\sigma}_{ij}^{x}, \hat{\sigma}_{ij}^{y}, \hat{\tau}_{ij}^{xy}, \hat{\sigma}_{ij}^{th}\right) \tag{4}$$

where β is a loading proportionality constant measuring the ratio of thermal to mechanical loads and is given by:

$$\beta = \frac{E_c \alpha_c \Delta T_c}{S} \tag{5}$$

In Equation 5 E_c, α_c and ΔT_c are the characteristic elastic modulus, Coefficient of Thermal Expansion (CTE) and characteristic temperature difference, while S is the magnitude of the proportional mechanical load measured in the units of stress.

Matrix Micro-Stresses

Figure 7 shows contour plots of stress induced in the plain weave C/SiC system due

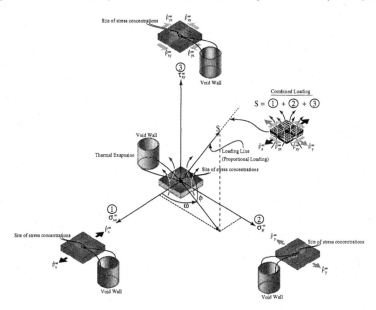

Figure 7: The induced matrix micro-stresses as a result of in-plane mechanical loads and free thermal expansion.

to in-plane mechanical loads and free thermal expansion. For the sake of clarity, a blow-up of stresses induced in the vicinity of the macroscopic void are shown adjacent to each contour plot. The average stresses in the void wall are calculated by first computing the stress at the center of each finite element belonging to the free surface adjacent to the void and then averaging the stresses in the z-direction, in the finite elements at the same angular distance θ from the center of the void. As a result, the induced stresses due to the action of mechanical tension along the x-direction are represented as shown in Figure 8. The $\hat{\sigma}_{xx}$ stress assumes maximum values at $\pm 90°$ from the center of the void. These profiles suggest that stress concentrations occur in the void along planes orthogonal to the direction of pull.

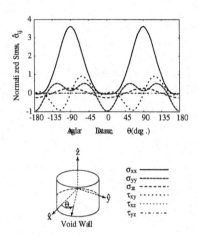

Figure 8: Profiles of induced matrix micro-stresses in the vicinity of the macroscopic void due to tension along the x-direction ($\omega = 0$, $\phi = 0°$).

The magnitudes of the stresses shown in Figure 8 suggest that the maximum induced non-dimensional $\hat{\sigma}_{xx}$ stress is of the order of about 3.5 times the remotely applied mechanical load. Similar results due to mechanical tension along the y-direction, pure remote xy-shear and thermal loads are presented elsewhere [15].

Overall the findings regarding the stress fields in woven CMCs are consistent with previous research efforts [5, 14]. In both works cited above, it was shown clearly that damage bands begin to emanate from the vicinity of macroscopic voids in woven CMCs. These damage bands were also shown to be the most likely sites of initiation of cracking in the matrix material. As such, the stress results obtained in the present work inspire confidence in the finite element model.

The total stress induced in the vicinity of the matrix void is calculated with the aid of Equation 4 for different values of β. The effects of β which at least theoretically can take on values from $-\infty$ to $+\infty$ are explored through extensive parametric studies reported elsewhere [15]. However, in this work we present only the results corresponding to two distinct values of $\beta = $ -3 and +3. Since the characteristic material and the remotely applied proportional mechanical load (S) do not change, each individual value of β really measures the total stress induced in the matrix void at a unique characteristic temperature difference. More specifically, $\beta \longrightarrow -\infty$ captures the case of pure thermal loading wherein $\Delta T_c < 0$ whereas $\beta \longrightarrow +\infty$ implies pure thermal loading with $\Delta T_c > 0$. As shown in Figure 9, with $\beta = -3$, $\hat{\sigma}_{xx}^{max}|_{\theta=\pm90°} = 5.0$ whereas with $\beta = +3$, $\hat{\sigma}_{xx}^{max}|_{\theta=\pm90°} = 2.2$, implying that stress concentrations increase as $\beta \longrightarrow -\infty$ and vice verse.

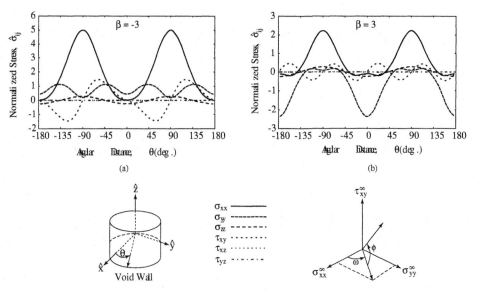

Figure 9: Profiles of total induced matrix micro-stresses in the vicinity of the macroscopic void, due to the combined effects of mechanical tension in the x-direction and thermal loading; i.e., $\omega = \phi = 0°$ and (a) $\beta = -3$ and (b) $\beta = +3$ with $\hat{\alpha}_f = \alpha_f/\alpha_m = 0.3$

Maximum Normal Stress Criterion

Using the functional relation of the total induced stress in the matrix void and the results shown in Figure 9, we will now develop a matrix failure function based on the normal stress criterion. According to normal stress criterion, a given material fails when the maximum normal stress at a point in the system reaches the ultimate tensile strength of the material. We can then write:

$$\sigma_{p1} = \sigma_{mf} \tag{6}$$

where σ_{p1} is the maximum normal stress and σ_{mf} is the matrix failure stress. We can obtain σ_{p1} as the biggest eigenvalue of the characteristic equation formed with the stress tensor given by σ_{ij}^{total}. In other words we solve the characteristic equation:

$$\left| \sigma_{ij}^{total} - \lambda \delta_{ij} \right| = 0 \tag{7}$$

where δ_{ij} is the Kronecker delta taking on a value of 1 if $i = j$ and 0 if $i \neq j$. The eigenvalues of σ_{ij}^{total} are obtained as the roots of the cubic characteristic equation of Equation 7 and σ_{p1} is assigned the value of the biggest eigenvalue as follows:

$$\sigma_{p1} = S \hat{\lambda}_1 \left(\hat{\Sigma}_{ij}^{total} \right) \tag{8}$$

where $\hat{\lambda}_1$ is biggest eigenvalue of the normalized stress tensor given by $\hat{\Sigma}_{ij}^{total}$. Normalizing Equation 8 with the proportional mechanical load S and invoking the normal stress criterion Equation 6, we have a functional form of matrix failure given by:

$$\frac{\sigma_{mf}}{S} = \Omega_1 \left(\omega, \phi, \beta, \hat{\sigma}_{ij}^x, \hat{\sigma}_{ij}^y, \hat{\tau}_{ij}^{xy}, \hat{\sigma}_{ij}^{th} \right) \tag{9}$$

where the arguments of Ω_1 have the same meaning as explained in Equation 4.

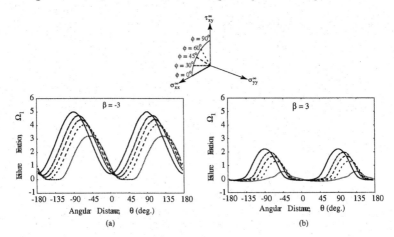

Figure 10: Profiles of the failure function based on the normal stress criterion. Combined effects of remote in-plane mechanical tension along the x-direction ($\omega = \phi = 0°$) and (a) $\beta = -3$ and (b) $\beta = +3$ with $\hat{\alpha}_f = \alpha_f / \alpha_m = 0.3$

Figure 10, reports on the effect of ω and β on Ω_1. To be consistent with the results presented for $\hat{\sigma}_{ij}^{total}$ in Figure 9, we present here results obtained for Ω_1 with $\beta = -3$ and $+3$ combined with mechanical loading in the x-direction ($\omega = 0°$). The five curves in both Figure 10 (a) and (b) correspond to increasing amounts of remotely applied in-plane mechanical shear load as measured via the spherical loading parameter ϕ. As expected, the failure function Ω_1 takes on maximum values when $\phi = 0°$ corresponding to zero in-plane shear load, regardless of the value of β. Another interesting feature represented by the results in Figure 10 is that the failure function Ω_1 takes on maximum values at $\pm 90°$ corresponding to $\phi = 0°$ and thereafter the magnitude of Ω_1 decreases with increasing shear load. Also it is worth noting that $|\Omega_1^{max}(\beta)|_{\phi=90°} \simeq |\Omega_1^{max}(\beta)|_{\phi=0°}/2$. These profiles of Ω_1 in Figure 10 suggest that failure is most likely to initiate at "equatorial" planes from the center of the matrix void regardless of the value of either β or ϕ.

Matrix Failure Loci

In order to further understand the effects of applied mechanical loads and temperature gradients on the matrix failure stress, we developed matrix failure loci for the current plain weave C/SiC system being studied. The proportional loading vector $S = \sigma_{xx}^{\infty}\hat{i} + \sigma_{yy}^{\infty}\hat{j} + \tau_{xy}^{\infty}\hat{k}$ was employed along with the loading proportionality constant β to develop profiles of the failure function Ω_1 as shown in Figure 10. The maxima $\Omega_1^{max}(\beta)$ corresponding to Ω_1 was then used to develop a failure point in the normalized $\hat{\sigma}_{xx}^{\infty} - \hat{\tau}_{xy}^{\infty}$ space, for different combinations of ω and ϕ. Therefore, for a given remotely applied mechanical load controlled by the spherical loading parameters ω and ϕ, the matrix would fail in accordance with the normal stress criterion if:

$$\frac{\sigma_{xx}^{\infty}}{\sigma_{mf}} > \frac{cos\phi cos\omega}{\Omega_1^{max}(\beta)}$$

$$\frac{\tau_{xy}^{\infty}}{\sigma_{mf}} > \frac{sin\phi}{\Omega_1^{max}(\beta)}$$

(10)

where $\Omega_1^{max}(\beta)$ implies that the maxima exhibited by Ω_1 is a function of the loading proportionality constant β. The matrix failure loci developed using this approach are shown in Figure 11.

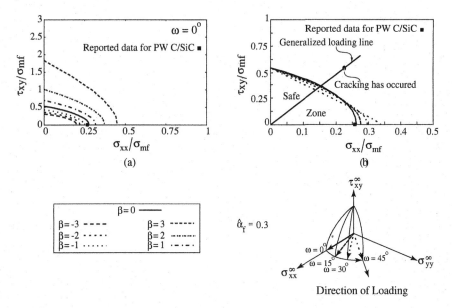

Figure 11: Profiles of the matrix failure loci for a plain weave C/SiC system based on the normal stress criterion.

Figure 11 (a) shows predicted failure for the plain weave C/SiC system under the combined action of remote in-plane mechanical tension in the x-direction and increasing amounts of remote in-plane xy-shear along with thermal loading. It should be noted that when $\omega = 0°$, $\sigma_{yy}^\infty = 0$ and when $\omega = 45°$, $\sigma_{xx}^\infty = \sigma_{yy}^\infty$. In the absence of any temperature gradient ($\beta = 0$), the failure loci corresponding to increasing amounts of biaxial tension are shown in Figure 11 (b). The trends in the normalized failure loci shown in both Figure 11 (a) and (b) indicate that initiation of matrix cracking requires higher applied pure shear loads in comparison to pure tension loads for any given β, which gives rise to elliptical failure loci. In fact, from the results presented in Figure 11 it is clear that it takes about one-half of the remotely applied shear load to initiate matrix cracking in pure tension.

The experimentally established proportional limit data used to compare model predictions were provided by [6] (see Table II). When this data was normalized with respect to the SiC matrix failure stress of $\sigma_{mf} \simeq 300$MPa (also given in Table II) and superimposed on the failure loci, an excellent agreement between model predictions and experimental data was achieved for the case when $\omega = \phi = 0°$ (pure remote tension) and $\beta = 0$ (no thermal temperature gradients) as shown in Figure 11 (b).

It is also evident from Figure 11 that as $\beta \longrightarrow -\infty$, matrix cracking is predicted to initiate at lower remotely applied mechanical loads. This result implies that at higher temperature changes (ΔT), the strength of woven CMCs decreases due to significantly increased amounts of thermal stresses owing to the large mismatch in the CTEs of the fibers and matrix material.

CONCLUSIONS

A robust modeling framework for assessing the strength of woven CMCs under remotely applied in-plane mechanical loads and thermal temperature gradients has been developed. The detailed geometry models of woven CMCs employed in this work capture the local stress distribution around the vicinity of macroscopic matrix voids very accurately. The stress concentrations around these matrix voids were shown to be 3 to 4 times greater (for $\beta = 0$) than the remotely applied mechanical loads. With increasing temperature changes the thermal stresses appear to play a greater role most likely leading to a degradation in the apparent high temperature first cracking strength of woven CMCs. The adopted modeling approach is robust and can be employed to assess the strength of satin weave CMCs such as the four harness, five harness and eight harness systems. However more substantive comparisons with experimentally established temperature sensitive failure data for woven CMCs are needed for model validation.

References

[1] B. N. Cox and G. Flanagan. Handbook of Analytical Methods for Textile Composites. Technical report, NASA, March 1997.

[2] J. Lamon, N. Lissart, C. Rechiniac, D.H. Roach, and J.M. Jouin. Micromechanical and Statistical Approach to the Behavior of CMCs. *Ceramic Engineering Science Proceedings*, 14:1115–1124, 1993.

[3] X. Aubard, J. Lamon, and O. Allix. Model of the Nonlinear Mechanical Behavior of 2D SiC-SiC Chemical Vapor Infiltration Composites. *Journal of the American Ceramic Society*, 77(8):2118–2126, 1994.

[4] R. N. Singh. Fracture and Crack Growth in Ceramic Matrix Composites at High Temperatures. *Presented at the 106th Annual Meeting & Exposition of the American Ceramic Society, Indianapolis, IN.*, 2004.

[5] S. I. Haan. *Modeling of the Mechanical Response of Plain Weave Composites*. PhD thesis, The University of Maryland, 2000.

[6] M. C. L. Patterson. Private Communication. Advanced Ceramics Research, Tucson, AZ. Formerly at Ceramic Composites Inc., Millersville, MD.

[7] J. L. Kuhn and P. G. Charalambides. Modeling of Plain Weave Fabric Composite Geometry. *Journal of Composite Materials*, Vol. 33(3):188–220, 1999.

[8] J. L. Kuhn and P. G. Charalambides. Elastic Response of Porous Matrix Plain Weave Fabric Composites:Part I-Modeling. *Journal of Composite Materials*, Vol. 32(16):1426–1471, 1998.

[9] J. D. Whitcomb and X. Tang. Effective Moduli of Woven Composites. *Journal of Composite Materials*, 35(23):2127–2144, 2000.

[10] J. A. Hewitt, D. Brown, and R. B. Clarke. Computer Modelling of Woven Composite Materials. *Composites*, Vol. 26(2):143–140, 1995.

[11] J. L. Kuhn. *Mechanical Behavior of Woven Ceramic Matrix Composites*. PhD thesis, The University of Maryland, 1998.

[12] J. L. Kuhn and P. G. Charalambides. Elastic Response of Porous Matrix Plain Weave Fabric Composites:Part II-Results. *Journal of Composite Materials*, Vol. 32(16):1472–1507, 1998.

[13] J. L. Kuhn, S. I. Haan, and P. G. Charalambides. A Semi-Analytical Method for the Calculation of the Elastic Micro-Fields in PLain Weave Fabric Composites Subjected to In-Plane Loading. *Journal of Composite Materials*, Vol. 33(3):221–266, 1999.

[14] J. L. Kuhn, S. I. Haan, and P. G. Charalambides. Stress Induced Matrix Microcracking in Brittle Matrix PLain Weave Fabric Composites Under Uniaxial Tension. *Journal of Composite Materials*, Vol. 34(19):1640–1664, 2000.

[15] M. P. Rao, M. Pantiuk, and P.G. Charalambides. A Proportional Limit Model for Establishing the Failure Loci of Plain and Satin Weave Ceramic Matrix Composites. *In progress*.

Author Index

Keyword Index